PREPARE YOU [barcode: D0196091]
THE IMPROBABLE

WHAT GROUNDBREAKING PHYSICS MIGHT EXPLAIN HOW CATS (ALMOST) ALWAYS MANAGE TO LAND ON THEIR FEET? (See page 54)

WHAT DID SCIENTISTS DISCOVER AFTER GIVING A PERSONALITY TEST TO SCOTTISH SHEEP? (See page 79)

WHY ARE BUSINESS MANAGERS WITH WIDE FACES MORE LIKELY TO MAKE A PROFIT? (See page 140)

DOES SEEING SOMEONE NUDE MESS UP YOUR MEMORY? (See page 94)

WHICH TOILET STALL GETS THE MOST ACTION IN ANTARCTICA? (See page 96)

HOW WOULD A PSYCHOANALYST INTERPRET YOUR STOMACH RUMBLES? (See page 102)

WHAT IS THE AIR CONDITIONING CAPACITY OF THE HUMAN NOSE? (See page 46)

WHY ARE BOOKS ON ETHICS MORE LIKELY TO GET STOLEN?* (See page 271)

DOES CARRYING A MOBILE PHONE AFFECT A RABBIT'S SEX LIFE? (See page 180)

---

* For the record, this is not a book on ethics.

THIS IS
IMPROBABLE

Cheese String Theory, Magnetic Chickens,
and Other WTF Research

Marc Abrahams

ONEWORLD

A Oneworld Book

First published by Oneworld Publications in 2012

This paperback edition published by Oneworld Publications 2013

Copyright © Marc Abrahams 2012

The moral right of Marc Abrahams to be
identified as the Author of this work has been
asserted by him in accordance with the Copyright,
Designs and Patents Act 1988

ISBN 978-1-85168-975-0
eISBN 978-1-78074-114-7

Typesetting and ebook design by Tetragon
Printed and bound in Great Britain by CPI Group (UK) Ltd, Croydon, CR0 4YY

Illustration credits on p.301

Oneworld Publications
10 Bloomsbury Street, London WC1B 3SR, England

Stay up to date with the latest books,
special offers, and exclusive content from
Oneworld with our monthly newsletter

Sign up on our website
www.oneworld-publications.com

CONTENTS

WITHOUT DUCKS

There are no homosexual, necrophiliac ducks in this book.* There wasn't room. Too many other improbable stories require space here.

It can be tempting to assume that 'improbable' implies more than that – implies bad or good, worthless or valuable, trivial or important. Something improbable can be any of those, or none of them, or all of them, in different ways. Something can be bad in some respects and good in others.

Improbable is, simply: what you don't expect.

I collect stories about improbable things, things that make people laugh, then think. The research, events, people, and pages in this book defy any quick attempt at judgement (bad-or-good? worthless-or-valuable? trivial-or-important?). But don't let that stop you from trying. See what you make of these:

> Measuring how cats skulk. Mechanically plucking and packaging a hijacker, then ejecting and delivering him by parachute to authorities on the ground below. Making people read inappropriately highlighted textbooks. Determining a person's natural hopping frequency. Watching volunteers as they listen to fingernails scraping on a blackboard. Monitoring the brain of a pianist as he repeats a short song non-stop for twenty-eight hours. Strolling with one shoelace untied, in country after country. Engineering a bra that can quickly convert into a pair of protective face masks. Placing a cat on a cow, then exploding paper bags every ten seconds. Optimizing the packaging for a large hollow chocolate bunny. Applying the thoughts of the French philosopher Foucault to the lives of Australian Rules football players. Adapting Jesus's strategic

leadership principles for the US Army. Plumbing the psyche of fruit machine gamblers. Classifying the kinds of boredom felt by mid-level administrators in the British Empire. Surgically altering a Belgian so he resembles singer Michael Jackson. Examining porcupine copulation. Discovering that electro-ejaculation is difficult to perform on the rhinoceros.

Most of what you'll read here first appeared, in some form, in my weekly 'Improbable Research' column in the *Guardian*. But the plod of science does go on, and so for this book I've dug up more details, added updates, and tossed in extra improbable titbits.

There's more to each of these stories than I could fit into these pages, of course. That's partly why I give you citations. Surprises await you, should you choose to follow those leads.

There's another, more demanding reason to include the citations. Some people think these stories are fictional or exaggerated. No, friend: these are nonfiction. I have tried hard to exaggerate nothing.

Sincerely and improbably,

Marc Abrahams

Editor and Co-founder,
Annals of Improbable Research

---

* Now, about those ducks. If you want to read about them, or find out what's happened on the several scientific research fronts that grew up in the wake of their discovery, you'll have to go look on your own. The place to start is Kees Moeliker's now-historic report 'The First Case of Homosexual Necrophilia in the Mallard *Anas platyrhynchos* (Aves: Anatidae)', published in 2001 in the Dutch biology research journal *Deinsea* (see volume 8, pages 243–47; no translator needed). Or take the special tour at the Natural History Museum in Rotterdam where Kees, the museum's curator, one afternoon (a) noticed a sudden loud sound that turned out to be a duck fatally crashing at high speed into the museum's glass wall, then (b) took up his notebook and camera to document how the event played out over the next seventy-five minutes.

Or you could skip ahead to watch videos of the 2003 Ig Nobel Prize ceremony at Harvard University, where Kees was awarded that year's Ig Nobel in the field of biology. Alternatively, you could read the book *De eendenman* (the title translates as *The Duck Guy*), which Kees wrote a few years later, after a publisher nagged him to do it; the book became so popular in the Netherlands that it was reprinted five times during the first two weeks. Or you could read some of the reports Kees subsequently wrote for the *Annals of Improbable Research*, for which he now serves as European Bureau Chief. His reports cover many topics. Birds that spend their days repeatedly hurling themselves against a particular window. Historic murders of sparrows by cricket players and television producers. International biomedical concern about the possible disappearance of pubic lice. And more.

Many of the people described in this book are like Kees Moeliker, in at least one way. When they give their attention to a particular question, they can be entertainingly focused about it.

THIS IS IMPROBABLE

ONE

STRANGE IN THE HEAD

IN BRIEF

'MY GRANDMOTHER'S PERSONALITY: A POSTHUMOUS EVALUATION'
by Frederick L. Coolidge (published in the *Journal of Clinical Geropsychology*, 1999)

Some of what's in this chapter: Thinking, to the brink of medical danger • Spotting hairy heads in theme parks • Being bored for His Majesty • Playing on, whilst being watched • Being brain damaged, for better wagering • Fingering beauty for intelligence • Heading up Brain • An Alias for body hair • Praying, to the brink of madness.

YOUR MIND COULD KILL YOU

Exactly how dangerous is it to think? The question matters, because for some people it truly is dangerous – physically, life-threateningly dangerous.

This question also bears on one that's seemingly unrelated: is it dangerous for students to use a calculator, rather than do maths in their heads?

In 1991, researchers in Osaka, Japan published a report called 'Reflex Epilepsy Induced by Calculation Using a "Soroban", a Japanese Traditional Calculator'. (The English word for *soroban* is 'abacus'.) The report describes an unfortunate young man who 'entered college in 1980, where he belonged to a music club and was in charge of the drums. After six months, he felt intense psychological tension during drum playing and particularly when he had to write musical scores phrase after phrase while listening to the music recorded on tape.' The situation worsened. Writing musical scores sometimes induced generalized tonic-clonic convulsions. The man truly suffered for his music.

In his final year at university, he discovered that doing calculations on an abacus caused the same problem, with even more severity. He stopped using an abacus, and started seeing doctors.

Specialists have seen and reported other such cases.

Consider the disturbingly thought-provoking paper entitled 'Seizures Induced by Thinking'. The report was published by A. J. Wilkins and three colleagues at the University of Essex in the *Annals of Neurology* in 1982. The researchers describe a man who suffered convulsions whenever he performed certain kinds of mental arithmetic. This was pure mental calculation, without the complication of an abacus or other mechanical or electronic apparatus. Mental addition seemed harmless enough for this man, and so did mental subtraction. But whenever he tried doing multiplication in his head, it triggered seizures. Division was equally a danger.

Other cases on record hint that subtraction is not always as safe as it seems, at least not for absolutely everyone. Nor is addition.

Mathematics and musical composition are not the only hazardous mental activities. A team at St Thomas' Hospital in London documented the plight of seventeen people who must watch what they watch. For them, the act of reading can trigger seizures. Newspapers are dangerous. Books are dangerous. Perilous materials are everywhere. There are also people for whom the act of writing is dangerous.

So, in reading, in writing, in arithmetic, and in other kinds of thought, true dangers lurk. They are exceedingly rare. At least, that's what the doctors say they think.

Wilkens, A. J., B. Zifkin, F. Andermann, and E. McGovern (1982). 'Seizures induced by thinking.' *Annals of Neurology* 11: 608–12.
Yamamoto, Junji, Isao Egawa, Shinobu Yamamoto, and Akira Shimizu (1991). 'Reflex Epilepsy Induced by Calculation Using a "Soroban", a Japanese Traditional Calculator.' *Epilepsia* 32: 39–43.
Koutroumanidis, M., M. J. Koepp, M. P. Richardson, C. Camfield, A. Agathonikou, S. Ried, A. Papadimitriou, G. T. Plant, J. S. Duncan, and C. P. Panayiotopoulos (1998). 'The Variants of Reading Epilepsy. A Clinical and Video-EEG Study of 17 Patients with Reading-Induced Seizures.' *Brain* 121: 1409–27.

COMBING THROUGH THE DATA

Clarence Robbins and Marjorie Gene Robbins visited theme parks hoping to find a good, representative mix of hairy-headed strangers. They then wrote 'Hair Length in Florida Theme Parks: An Approximation of Hair Length in the United States of America'. The study tells how Robbins and Robbins gathered their data, combed through it, and extrapolated the strands to gain a new understanding of America.

At the time of their investigation, Robbins and Robbins were the leading researchers at Clarence Robbins Technical Consulting, a think tank located in their home in Clermont, Florida. Clermont is just a short drive from four big theme parks – Epcot, Universal Studios, the Magic Kingdom, and MGM Studios. In visiting those parks, the researchers set themselves a simple, clear goal: 'to obtain data on the percentage of persons in the US with different lengths of scalp hair'.

The goal was not so easily attained. Robbins and Robbins found it prudent to make two additional theme park visits specifically to address questions pertaining to accuracy.

The first extra visit was to 'determine whether or not any hairstyles might interfere with or affect our estimates on free-hanging hair length'. This proved susceptible to an easy statistical adjustment.

The other visit was to 'determine whether or not any headcovers' – by which they meant caps, hats, and scarves – would skew the estimates. They decided, happily, that headcovers cause no such problems.

Robbins and Robbins could not, of course, ensure that their hairy-headed sample accurately represented the entire American populace. But the monograph tells how they tried: 'In an attempt to try to determine how this population relates to the general US population, several telephone calls were made to the Walt Disney Corporation, including their Market Research Department. Those contacted refused to provide any helpful information, indicating that their data and results were proprietary.'

The Robbins–Robbins study, though technical in nature, also presents facts that may be enlightening to laypersons: 'One woman who was observed at Epcot had hair reaching several centimeters past her buttocks. She was dressed in a skintight costume, as were two young men walking with her shortly after a Disney parade. She had curly blonde hair and appeared to be in her mid to late 20s. This woman was most likely a Disney employee, hired for her long hair, because we observed her once before in a Disney parade playing Rapunzel.'

The report, published in the *Journal of Cosmetic Science*, concludes with a compelling summary: 'By observing the hair of 24,300 adults in central Florida theme parks at specified dates from January through May of 2001 and estimating hair length relative to specific anatomical positions, we conclude that about 13% of the US adult population currently has hair shoulder-length or longer, about 2.4% have hair reaching to the bottom of the shoulder blades or longer, about 0.3% have hair waist-length or longer, and only about 0.017% have hair buttocks-length or longer. Hair appreciably longer than buttock-length was not observed in this population.'

Robbins, Clarence, and Marjorie Gene Robbins (2003). 'Scalp Hair Length. I. Hair Length in Florida Theme Parks: An Approximation of Hair Length in the United States of America.' *Journal of Cosmetic Science* 54 (1): 53–62.

TRUTH ON THE SIDE

On which side lies the truth? It's on the left, according to a 1993 study published in the journal *Neuropsychologia*. The left ear, this study says, is better than the right ear at discerning truth. Slightly better. In most people. Some of the time.

The experiment, called 'Hemispheric Asymmetry for the Auditory Recognition of True and False Statements', was conducted by Franco Fabbro and his team at the University of Trieste. Twenty-four men and twenty-four women each donned headphones, and then (presumably) followed these instructions: 'In the headphones you will hear phrases pronounced by four people you don't know. There are two types of phrase – "This is a pleasant photo" and "This is an unpleasant photo". While they pronounced these phrases they were looking at photographs which they had previously judged to be pleasant or unpleasant. Sometimes they are telling the truth. Sometimes they are lying. After hearing a phrase you have to decide whether you think the speakers are telling the truth or a lie.'

The effect is subtle. According to the data, the left-ear truth detector is not especially good at recognizing women's lies. It works, to the extent it works, only when a man does the lying. Even then, it correctly recognizes only sixty-three percent of the true statements as being true.

Fabbro and his colleagues were intrigued by the two cerebral hemispheres. One is thought to be more skilled than the other at handling emotions. 'Most people undergo an increase in emotional stress when telling a lie', the study says. The theory, too, is subtle. Fabbro and his colleagues phrase it this way: 'Since in human cultures lying prevailingly occurs at the verbal level, it is reasonable to expect a stronger tendency to consider false that kind of information which is transmitted and processed through verbal systems. For the same token it is reasonable to expect a stronger tendency to consider true the information which is processed by non-verbal systems.'

A different experiment, conducted more than a decade earlier at McGill University in Montreal, Canada, looked for something less subtle. And found it.

Walter W. Surwillo told all in his study called 'Ear Asymmetry in Telephone-Listening Behavior', which was published in the journal *Cortex*. He, too, was curious about the powers of the left ear versus the right. Surwillo surveyed people whose jobs involve a lot of telephoning, and also people whose jobs don't. His question: which ear do they prefer for listening on a telephone?

The results: 'Listening with the left ear was associated with heavy use of the telephone. The most frequently given reason for listening with the left ear was that it freed the right hand for writing and dialing. This preference would appear to be motivated by convenience for although either ear is available for listening, it is easier to hold the receiver to the left than the right ear while grasping it in the left hand.'

For further elucidation of the relative power of the left and right ears, you may wish to consult a study of 203 telesales staff examining the relationship between ear preference, personality, and job performance ratings. Those who preferred a right ear headset generally rated higher in supervisors' eyes when it came to actual performance and developmental potential compared to those staff who preferred a left ear headset.

Fabbro, F. B., B. Gran, and A. Bava (1993). 'Hemispheric Asymmetry for the Auditory Recognition of True and False Statements.' *Neuropsychologia* (31) 8: 865–70.

Surwillo, Walter W. (1981). 'Ear Asymmetry in Telephone-Listening Behavior.' *Cortex* 17 (4): 625–32.

Jackson, Chris J., Adrian Furnham, and Tony Miller (2001). 'Moderating Effect of Ear Preference on Personality in the Prediction of Sales Performance.' *Laterality: Asymmetries of Body, Brain, and Cognition* 6 (2): 133–40.

IMPERIAL BOREDOM

True, at its height, the British Empire produced magnificent heaps of wealth and power. But according to historian Jeffrey Auerbach, the empire also generated staggering amounts of boredom.

In a copiously documented report in the journal *Common Knowledge*, Auerbach writes: 'Throughout the nineteenth century and into the twentieth, British imperial administrators at all levels were bored by their experience traveling and working in the service of king or queen and country. Yet in the public mind, the British empire was thrilling – full of novelty, danger, and adventure, as explorers, missionaries, and settlers sailed the globe in search of new lands, potential converts, and untold riches.'

Auerbach's interests are not limited to boredom. An assistant professor of history at California State University, Northridge, he has published on many other subjects, among them 'The Homogenization of Empire', 'The Monotony of Empire', and the inspirationally titled 'The Impossibility of Artistic Escape'.

The imperial boredom report is filled with telltale evidence of administrators' boredom. Those administrators range from the soon-to-be celebrated Winston Churchill (who at age twenty-one wrote that Indian life was 'dull and uninteresting') to the clerk who wrote this ditty:

> From ten to eleven, ate a breakfast at seven;
> From eleven to noon, to begin 'twas too soon;
> From twelve to one, asked 'What's to be done?'
> From one to two, found nothing to do;
> From two to three, began to foresee
> That from three to four would be a damned bore.

Auerbach complains that, for generations, 'Scholars have by and large perpetuated [a] glamorous view of the empire, portraying imperial men either as heroic adventurers who charted new lands and carried "the white man's burden" to the farthest reaches of the planet or as aggressors who imposed culturally bound norms and values on indigenous peoples and their ways of life.'

He says he did his research by 'reading against the grain of published memoirs and travel logs' and by digging into the unspectacular mutterings of private diaries and letters. His task was the more difficult, he argues, because 'if people felt bored before the mid-eighteenth century, they did not know it'. This view of boredom, he points out, was persuasively developed by Patricia Meyer Spacks, whose 304-page book *Boredom: The Literary History of a State of Mind*, titillated thrill-starved scholars in 1995.

Auerbach is himself writing a book. It's about imperial boredom. He will need to expand considerably upon his current study, which is twenty-three pages long and includes only seventy-six footnotes. As this book – the one you are reading goes to press, Auerbach's book has not yet hit the market. His website lists it as *Imperial Boredom: The Banality of the British Empire, 1757–1939* (in progress).

Perhaps, though, Auerbach has already done his finest work. Here, in just forty-seven words, is his take on imperial boredom: 'The reality simply could not live up to the expectations created by newspapers, novels, travel books, and propaganda. As a consequence, notwithstanding some famous exceptions, nineteenth-century colonial officials were deflated by the dreariness of their imperial lives, desperate to ignore or escape the empire they had built.'

Auerbach, Jeffrey (2005). 'Imperial Boredom.' *Common Knowledge* 11 (2): 283–305.

28 STRAIGHT HOURS OF VEXATIONS

German and Austrian researchers analysed what happened to pianist Armin Fuchs when he spent more than a full day playing over and over again, nonstop, an oddly named piece of music by a French composer. They also analysed what happened to the music. This was a tour de force of artistic and neurological repetition.

The research team – Christine Kohlmetz, Reinhard Kopiez, and Marc Bangert of the Hanover University of Music and Drama, and Werner Goebl and Eckart Altenmüller of the Austrian Research Institute for Artificial Intelligence, in Vienna – published a pair of monographs in 2003 describing what they measured in the pianist.

The study titles, like the performance, are lengthy. One, in the journal *Psychology of Music*, includes the phrase 'Electrocortical Activity in a Pianist Playing "Vexations" by Erik Satie Continuously for 28 Hours'. When Satie composed the piece in 1893, he added an instruction that the performer should: 'play this motif 840 times in succession'.

Twenty-first-century researchers Kohlmetz, Kopiez, Bangert, Goebl, and Altenmüller explicitly question whether nineteenth-century composer Satie 'was aware of the effect his work might have on a pianist, especially with regard to his or her consciousness and motor function'.

Do the maths and you'll see that Fuchs, the intrepid piano player, averaged thirty performances per hour. That's a lot of 'Vexations', in two-minute-long chunks.

Time	d [hr]	Protocol of spectator	Protocol of pianist (AF)	State
9 am–5 pm	−8.00		Got up, breakfast, preparation in concert hall, interview, setting electrodes for EEG, kinesiologic exercises	
5 pm	0.00	**Beginning of performance**	No chance to concentrate before the beginning	**Alertness**
6 pm	1.00			0.10–2.10 hr
7 pm	2.00		Relaxed state of mind	
8 pm	3.00			
9 pm	4.00		First complaints (pain in shoulder, clavicula)	
10 pm–3 am	5.00–10.00			
4 am	11.00	First meal, first cigarette	Extremely tired, hard to concentrate	
	11.25	**Signs of beginning trance**	Better after smoking	
5 am	12.00			
6 am	13.30	2nd cigarette, mixes up passages, stops to measure	Approx. 6.30 am **beginning of trance**, felt extremely tired,	

Detail from table: 'Event protocol of the performance recorded by an observer and retrospectively by the pianist'

By affixing wires to Fuchs's head and using an electroencephalograph (EEG) to continuously monitor gross electrical activity in his brain, the researchers discovered that 'the pianist experienced different states of consciousness throughout the performance ranging from alertness to trance and drowsiness'.

Fuchs's playing grew inconsistent during the periods of drowsiness. But when alert, the man was a model of consistency. 'Most importantly', says the study, 'whilst in deep trance, which included effects such as time-shortening, altered perception and characteristic changes in the EEG, the pianist managed not only to keep on playing but also to maintain a constant tempo, hence executing complex motor schemes at a high level of performance.' (For a video of the performance in progress, visit http://www.youtube.com/watch?v=km9GiejF5OQ.)

The second study, published in the *Journal of New Music Research*, gives a much finer-grained 'tempo and loudness analysis'. It boasts that 'non-linear methods revealed that changes in loudness and tempo are of a highly complex nature, and both parameters unfold in an 18-dimensional space. This has never before been demonstrated in performance research.'

'Vexations' is not one of Erik Satie's most beloved compositions, not yet anyway. Although brief (when played, contrary to instructions, just a single time), it meanders along in a way that's neither quick nor catchy.

But Satie pioneered something of value – the technique ('play this motif 840 times') that the music business would refine many years later, with significant payoff. Radio disc jockeys demonstrated in the 1950s that, by persistently, diligently repeating a song, they could ensconce it into the minds of many listeners, who would ever after believe it to be one of their favourite tunes.

Kohlmetz, Christine, Reinhard Kopiez, and Eckart Altenmüller (2003). 'Stability of Motor Programs during a State of Meditation: Electrocortical Activity in a Pianist Playing "Vexations" by Erik Satie Continuously for 28 Hours.' *Psychology of Music* 31 (2): 173–86.
Kopiez, Reinhard, Marc Bangert, Werner Goebl, and Eckart Altenmüller (2003). 'Tempo and Loudness Analysis of a Continuous 28-Hour Performance of Erik Satie's Composition "Vexations".' *Journal of New Music Research* 32 (3): 243–58.

THE MIND OF THE WAITER

Strategies of Buenos Aires waiters to enhance memory capacity in a real-life setting

Tristan A. Bekinschtein[a,b], Julian Cardozo[a] and Facundo F. Manes[a,c*]
[a]Institute of Cognitive Neurology (INECO), Buenos Aires, Argentina
[b]MRC Cognition and Brain Sciences Unit, Cambridge, UK
[c]Institute of Neurosciences - Favaloro University, Buenos Aires, Argentina

Abstract. Human learning and memory evaluation in real-life situations remains difficult due to uncontrolled variables. Buenos Aires waiters, who memorize all the orders without written support, were evaluated in situ. Waiters received either eight different

Buenos Aires boasts impressive waiters, whose minds are worth studying, according to the paper 'Strategies of Buenos Aires Waiters to Enhance Memory Capacity in a Real-Life Setting', published in the journal *Behavioural Neurology*. 'Typical Buenos Aires senior waiters memorize all the orders from respective clients and take the orders without written support of as many as ten persons per table', it explains. 'They also deliver the order to each and every one of the customers who ordered it without asking or checking.'

And most of the time, they get it right.

How do they manage it? Researchers Tristan Bekinschtein, Julian Cardozo, and Facundo Manes ran an experiment to find out. The three are based variously at the Institute of Cognitive Neurology and at Favaloro University, both in Buenos Aires, and at the MRC Cognition and Brain Sciences Unit of the University of Cambridge.

Eight customers sat around a table and ordered drinks. When the waiter returned with the beverages, the scientists tallied up how many were served to the people who had ordered them, and how many were mistakenly delivered to someone else. All the waiters performed admirably.

The customers later ordered more drinks, then switched seats before the waiter returned. This produced dreary results. The scientists tried this on nine waiters, only one of whom consistently delivered drinks to the right people.

Interviewed afterwards, the waiters explained that they generally paid attention to customers' locations, faces, and clothing. They also disclosed a tiny trick of the trade. They 'did not pay attention to any customer after taking a table's order, as if they were protecting the memory formation in the path from the table to the bartender or kitchen'.

In preparing their study, Bekinschtein, Cardozo, and Manes discovered a published account of a most remarkable waiter (they do not specify, though, whether he was Argentinian). This man had trained himself to almost Olympian awesomeness in the delivery of food. He 'could recall as many as 20 dinner orders, categorize the food (meat or starch) and link it to the location in the table. He also used acronyms and words to encode salad dressing and visualized cooking temperature for each customer's meat and linked it to the position on the table.'

The Buenos Aires waiters, in contrast, 'reported systematically that they have not thought of any particular strategy and that their great ability comes only with time and practice'. True or not, this answer accords with their profession's famed tradition of haughty disdain for theory.

The best waiter – the one who delivered drinks correctly even when customers had swapped seats – claimed that, unlike his colleagues, he ignored where customers sat, and paid attention only to their looks. His professional experience, he said, 'had been mostly in cocktail parties for 10 years, where people tend to change their position in the room; only in the last three years he had been working in the restaurant'.

Bekinschtein, Tristan A., Julian Cardozo, and Facundo F. Manes. 'Strategies of Buenos Aires Waiters to Enhance Memory Capacity in a Real-Life Setting.' *Behavioural Neurology* 20. 65–70.

GAMBLING WITH BRAIN DAMAGE

Brain damage can sometimes give gamblers a winning edge, an American study suggests. The researchers take a flier at explaining how and why certain brain lesions might, in some circumstances, help a person triumph over others or over adversity.

The study, published in the journal *Psychological Science*, renders its tantalizing, juicy question into lofty academese. The five co-authors, led by Baba Shiv, a marketing professor at Stanford University's Graduate School of Business, ask 'Can dysfunction in neural systems subserving emotion lead, under certain circumstances, to more advantageous decisions?'

The team experimented with people who had abnormalities in particular brain regions – the amygdala, the orbitofrontal cortex, and the right insular or somatosensory cortex. Medically, abnormalities in these areas can signal that something is amiss in how the person handles emotions.

Each brain-damaged person got a wad of play money, and instructions to gamble on twenty rounds of coin tossing (heads-you-win/tails-you-lose, with some added twists). Other people, who had no such brain lesions, got the same amount of money, and the same gambling instructions.

The brain-damaged gamblers pretty consistently ended up with more money than their healthier-brained competitors. The researchers speculate that when 'normal' gamblers encounter a run of unhappy coin-toss results, they get discouraged and become cautious – perhaps too cautious. Not so the people with brain-lesion-induced emotional dysfunction. Encountering a run of bad luck, they plunge on, undaunted. And then enjoy a relatively handsome payoff. At least sometimes.

The study notes that this brain damage side-benefit might occasionally even save someone's life. They cite the case of a man with ventromedial prefrontal damage who was driving under hazardous road conditions: 'When other drivers reached an icy patch, they hit their brakes in panic, causing their vehicles to skid out of control, but the patient crossed the icy patch unperturbed, gently pulling away from a tailspin and driving ahead safely. The patient remembered the fact that not hitting the brakes was the appropriate behavior, and his lack of fear allowed him to perform optimally.'

Lead author Baba Shiv has an eye for nonstandard ways of exploring the weird morass that is human behaviour. He sometimes teaches a course called 'The Frinky Science of the Mind'. And in 2008, he and three colleagues were awarded an Ig Nobel Prize for demonstrating that expensive fake medicine is more effective than cheap fake medicine.

Shiv, Baba, George Loewenstein, Antoine Bechara, Hanna Damasio, and Antonio R. Damasio (2005). 'Investment Behavior and the Negative Side of Emotion.' *Psychological Science* 16 (6): 435–39.

HEELS GIVE RISE TO SCHIZOPHRENIA

Do shoes cause schizophrenia? Jarl Flensmark of Malmo, Sweden, wants to know, and in a recent paper in the journal *Medical Hypotheses*, he explains why.

'Heeled footwear', he writes, 'began to be used more than 1000 years ago, and led to the occurrence of the first cases of schizophrenia.... Industrialization of shoe production increased schizophrenia prevalence. Mechanization of the production started in Massachusetts, spread from there to England and Germany, and then to the rest of Western Europe. A remarkable increase in schizophrenia prevalence followed the same pattern.'

The story, if accurate and true, is disturbing. Flensmark sketches the details.

'The oldest depiction of a heeled shoe comes from Mesopotamia, and in this part of the world we also find the first institutions making provisions for mental disorders,' he writes. 'In the beginning schizophrenia appears to be more common in the upper classes. Possible early victims were King Richard II and Henry VI of England, his grandfather Charles VI of France, his mother Jeanne de Bourbon, and his uncle Louis II de Bourbon, Erik XIV of Sweden, Juana of Castile [and] her grandmother Isabella of Portugal.' All of these individuals are either known or suspected of wearing heeled shoes.

He cites evidence from other parts of the world, too – Turkey, Taiwan, the Balkans, Ireland, Italy, Ghana, Greenland, the Caribbean, and elsewhere.

'Probably the upper classes began using heeled footwear earlier than the lower classes,' Flensmark points out. He then cites studies from India and elsewhere, which seem to confirm that 'schizophrenia first affects the upper classes'.

From these two streams of evidence – the rise of heels and the increase in documented cases of what may have been schizophrenia, Flensmark divines a strong connection. He modestly implies that he is not the first to do so. In the year 1740, he writes, 'the Danish-French anatomist Jakob Winslow warned against the wearing of heeled shoes, expecting it to be the cause of certain infirmities which appear not to have any relation to it'.

Flensmark boils the matter into a damning statement: 'After heeled shoes is [sic] introduced into a population the first cases of schizophrenia appear and then the increase in prevalence of schizophrenia follows the increase in use of heeled shoes with some delay.'

'I have,' he continues, 'not been able to find any contradictory data.'

Lest critics dismiss this as mere hand-waving or foot-tapping, Flensmark explains, biomedically, how the one probably causes the other: 'During walking synchronised stimuli from mechanoreceptors in the lower extremities increase activity in cerebellothalamocortico-cerebellar loops through their action on NMDA-receptors. Using heeled shoes leads to weaker stimulation of the loops. Reduced cortical activity changes dopaminergic which involves the basal gangliathalamo-cortical-nigro-basal ganglia loops.'

One could conclude that the medical establishment enjoys Flensmark's discovery. Virtually no one has stepped up to dispute it.

Flensmark, Jarl (2004). 'Is There an Association Between the Use of Heeled Footwear and Schizophrenia?' *Medical Hypotheses* 63 (4): 740–47.

PRETTY, CLEVER

Why beautiful people are more intelligent

Satoshi Kanazawa[a,*], Jody L. Kovar[b]

[a] *Interdisciplinary Institute of Management, London School of Economics and Political Science, Houghton Street, London WC2A 2AE, UK*
[b] *Department of Sociology, Indiana University of Pennsylvania, USA*

Abstract

Empirical studies demonstrate that individuals perceive physically attractive others to be more intelligent than physically unattractive others. While most researchers dismiss this perception as a "bias" or "stereotype," we contend that individuals have this perception because beautiful people indeed *are* more intelligent. The conclusion

Are beautiful people more intelligent than the rest of us? Satoshi Kanazawa and Jody Kovar think so. In their seventeen-page study 'Why Beautiful People Are More Intelligent', they explain bluntly: 'Individuals perceive physically attractive others to be more intelligent than physically unattractive others. While most researchers dismiss this perception as a "bias" or "stereotype", we contend that individuals have this perception because beautiful people indeed are more intelligent.'

Kanazawa, a reader in management and research methodology at the London School of Economics and Political Science, has become a brainy specialist on beauty: he entitled a subsequent study 'Beautiful Parents Have More Daughters'. Kovar is affiliated with Indiana University of Pennsylvania.

The pair of them apply detective skills and high intellect to some common beliefs: 'Critics have noted that people have the opposite stereotype that extremely attractive women are unintelligent. We do not believe such a stereotype exists, however. We instead believe that the stereotype is that blonde women and women with large breasts are unintelligent, both of which, just like the stereotype that beautiful people are intelligent, may statistically be true.'

Kanazawa and Kovar don't merely say these things. They back them up. The volume of their evidence, if not the evidence itself, is overwhelming. Nearly all of it comes from studies – lots of them – done by other people. Among the earlier discoveries:

> QUOTE: Middle-class girls ... have higher IQs and are physically more attractive than working-class girls.

> QUOTE: More beautiful children and adults of both sexes have greater intelligence (and thus) the maxim 'beauty is skin deep' is a 'myth'.

> QUOTE: Physical attractiveness has a significant positive effect on family income, personal income and education.

Yet, for all this, there is still hope for the physically bland. Kanazawa and Kovar explain there's a good possibility that any particular individual is not a dope: 'Our contention that beautiful people are more intelligent is purely scientific (logical and empirical); it is not a prescription for how to treat or judge others.... At the same time, our theory is probabilistic, not deterministic, and the available evidence suggests that the empirical correlation between physical attractiveness and intelligence ... is modest at best. Thus, any attempt to infer people's intelligence and competence from their physical appearance, in lieu of a standardized IQ test, would be highly inefficient.'

At the very end of their report, the two scientists suggest that beautiful people are more than just clever. The chain of logic, and its ultimate conclusions, are provocative:

1) Aggressive men are the most likely to achieve high status and to mate with beautiful babes;

2) Aggressive men are the most likely to have aggressive children, and beautiful babes are the most likely to have *beautiful* babies.

Add these together, Kanazawa and Kovar say, and you must conclude that, compared to everyone else, 'beautiful people are more aggressive'.

Kanazawa, Satoshi, and Jody L. Kovar (2004). 'Why Beautiful People Are More Intelligent.' *Intelligence* 32: 227–43.
Kanazawa, Satoshi (2007). 'Beautiful Parents Have More Daughters: A Further Implication of the Generalized Trivers-Willard Hypothesis (gTWH).' *Journal of Theoretical Biology* 244 (1): 133–40.

BRAIN ON HEAD IN BRAIN

Russell Brain – who was also Lord Brain, Baron Brain of Eynsham – was editor of the journal *Brain*. In 2011, the journal *Brain* celebrated the golden jubilee of the publication of Dr Brain's essay 'Henry Head: A Man and His Ideas'. Head preceded Brain (the man) as head (which is to say, editor) of the journal (the name of which, I note again for clarity, is: *Brain*).

Head headed *Brain* from 1905 until 1923. Brain became head in 1954, dying in office in 1967. No other editors in the journal's long history (it was founded in 1879) could or did boast surnames that so stunningly announced their obsession, profession, and place of employ. One of Dr Brain's final articles, in 1963, is called 'Some Reflections on Brain and Mind'.

Dr Head wrote many monographs, some quite lengthy, for *Brain*. The first, a 135-page behemoth, appeared in 1893, long before he became editor. In it, Dr Head gives special thanks to a Dr Buzzard, citing Dr Buzzard's generosity, the nature of which is not specified.

Reading Dr Brain's *Brain* tribute and other material about Dr Head, one gets the strong impression that Head had a big head, and that it was stuffed full of knowledge, which Dr Head was not shy about sharing. Brain writes that 'Some men ... feel impelled to impart information to others. Head was one of those.'

Brain then quotes Professor H. M. Turnbull as saying: 'I had the good fortune when first going to the hospital to meet daily in the mornings on the steam engine underground railway Dr Henry Head. He ... kindly taught me throughout our journeys about physical signs, much to the annoyance of our fellow travellers; indeed in his characteristic keenness he spoke so loudly that as we walked to the hospital from St. Mary's station people on the other side of the wide Whitechapel Road would turn to look at us.'

Brain says that Head 'would illustrate his lectures by himself reproducing the involuntary movements or postures produced by nervous disease, and "Henry Head doing gaits" was a perennial attraction'.

In 1904, at the age of forty-two, Head married a headmistress – Ruth Mayhew of Brighton High School for girls. Brain assures us that she was 'a fit companion for him in intelligence'.

Brain, though respectful of Head, suggests that his predecessor was over-brainy: 'He had many ideas; he bubbled over with them, and perhaps he was sometimes too ready to convince himself of their truth.'

Brain, Russell (1961). 'Henry Head: The Man and His Ideas.' *Brain* 84 (4): 561–66.
— (1963). 'Some Reflections on Brain and Mind.' *Brain* 86 (3): 381–402.
Head, Henry (1893). 'On Disturbances of Sensation with Especial Reference to the Pain of Visceral Disease.' *Brain* 16 (1-2): 1–133.
— (1923). 'Speech and Cerebral Localization.' *Brain* 46 (4): 355–528.
Rivers, W. H. R., and Henry Head (1908). 'A Human Experiment in Nerve Division.' *Brain* 31 (3): 323–450.

HAIR ON HEADS WITH BRAINS

Henry Head is not the only outstanding head in science. The Luxuriant Flowing Hair Club for Scientists is, as the name implies, a club for scientists who have luxuriant flowing hair. LFHCfS, as it is known unpronounceably to its members and their admirers, was founded in early 2001. Anyone can join, provided only that she or he is a scientist and has luxuriant flowing hair. And is proud of it.

The 'proud' part is important. The club is not for the morbidly shy, people-aversive scientist of stereotype and legend. Every LFH-CfS member's hair is on display on the club's website, at http://www.improbable.com/projects/hair/hair-club-top.html.

LFHCfS was founded by admirers of the famously curly mane of psychologist Steven Pinker. Pinker, then a professor at the Massachusetts Institute of Technology and now head of the psychology department at Harvard, enlisted as the first member. He proudly lists the club on his academic web page.

The membership ranks now include mathematicians, astronomers, linguists, organic chemists, computer researchers, immunologists, geneticists, physicists, neuroscientists, three sisters, a married couple, and other men and women of science, of both sexes and all hair colours and many hairstyles. There is even a real rock star, Italian chemist Piero

Paravidino, a guitar player for the heavy metal band Mesmerize and co-author of the paper 'Synthesis of Medium-Sized N-Heterocycles Through RCM of Fischer-Type Hydrazino Carbene Complexes'. Paravidino was named 2002 LFHCfS Man of the Year, beating out fellow LFH-CfS member and fellow rock star, the astronomer Brian May of Queen.

CALL FOR MEMBERS

Since the founding of the LFHCfS, scientists who once had long hair – but who no longer have hair – wrote asking if they could still be recognized, in some way. Thus was formed the LFHCfS's two sibling clubs: the Luxuriant Former Hair Club for Scientists, and the Luxuriant Facial Hair Club for Scientists. The members of all three clubs share a quality of extremeness. Each member has extremely good hair, or extremely none.

To propose yourself for membership, please send by email:
* A photograph that clearly provides evidence of (1) luxuriant, flowing head of hair; (2) luxuriant, flowing face of hair; or (3) luxuriant head of former hair.
* A current curriculum vitae outlining your credentials as a scientist.
* A pithy statement regarding your qualifications, both hirsute and scholarly.

You may nominate someone other than yourself, provided that the person meets the relevant membership qualifications and that you have first obtained that person's enthusiastic consent.

Send to marca@improbable.com with the subject line:
LUXURIANT HAIR CLUB

DR ALIAS, HAIR MAN

If a hairy man becomes insatiably curious about what it means to have all that hair, he may well run across the work of Dr A. G. Alias. Yes, that is his name.

Alias is an expert on certain aspects and implications of the hairiness of men. He has taken a special interest in hairy military leaders, hairy intelligentsia, low-ranked hairy boxers, and Marlon Brando. Alias has offered this shorthand version of his views: 'I am fairly certain that the vast majority of hairy/hirsute men, compared to the respective "much less" hirsute men of the same race and ethnic group, are strikingly more intelligent and/or educated, but only from a statistical point of view.'

Male hairiness enjoys a complex and often unclear relationship with intelligence and behaviour. Alias, who is based at the Chester Mental Health Center in Chester, Illinois, has tried to tease out a few of the many subtleties.

Alias attracted attention in 1996 when he presented a research paper called 'A Statistical Association Between Liberal Body Hair Growth and Intelligence' at the Eighth Congress of the Association of European Psychiatrists, held in London. He reported that hairiness is common among successful male academics, engineers, and physicians – and also among the men who join Mensa, the social club for high-scoring intelligence-test takers.

This was just a year after Alias had published a paper titled 'Top Ranked Boxers Are Less Hirsute Than Lower Level Boxers'. In it, he discusses mesomorphs – big-boned, muscular men. Alias carefully analysed 380 drawings in William Sheldon's book *Atlas of Men*. This was to gain a general understanding of whether big brutes have lots of body hair. Alias then carefully examined Harry Mullen's tome, *Great Book of Boxing*, in which 'body hair-revealing pictures are printed of 49 top-ranked, white heavyweight boxers, 15 of whom became world champions'. Alias concluded that, as a rule, champions were less hairy than non-champions. However, he cautions that the difference is not statistically significant.

Alias did not limit his research to white pugilists in dusty books; he also looked at black boxers who appeared on television. He reports that 'around 35% of the black boxers appear to be more hirsute than any of the 16 black boxers featured in [the book] *All-Time Greats of Boxing*: Johnson, Louis, Walcott, Patterson, Liston, Ali, Frazier, Holmes, Tyson, M. Spinks, Robinson, Hagler, Armstrong, Hearns, Leonard, and Saddler. Archie Moore, Ezzard Charles, Mike Weaver, Tony Tubbs, Iran Ian Barkley, and Lennox Lewis, who were conspicuously hirsute, heavyweight champions, were less than outstanding.'

That same year, 1995, Alias also released a paper called 'Non-

Pathological Associational Loosening of Marlon Brando: A Sign of Hypoarousal?' Biographers of the late actor can plumb it for unexpected insights.

Alias, A. G. (1996) 'A Statistical Association Between Liberal Body Hair Growth and Intelligence.' Presented at the Eighth Congress of the Association of European Psychiatrists, London, UK, 12 July.

— (1995). 'Top Ranked Boxers Are Less Hirsute Than Lower Level Boxers: An Example For the Importance of 5-Alpha-Reductase.' *Biological Psychiatry* 37 (9): 612–13.

— (1995). 'Non-Pathological Associational Loosening of Marlon Brando: A Sign of Hypoarousal?' *Biological Psychiatry* 37 (9): 613.

AN IMPROBABLE INNOVATION

'METHOD OF CONCEALING PARTIAL BALDNESS'

a/k/a 'the comb-over' by Frank J. Smith and Donald J. Smith (US Patent no. 4,022,227, granted 1977 and honoured with the 2004 Ig Nobel Prize in engineering)

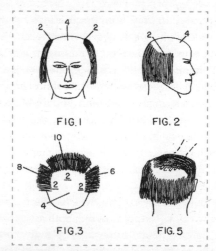

Selected steps from 'Method of Concealing Partial Baldness'

BALDLY ACCOUNTABLE

It's your own fault if you go bald, or if you lose your memory, or both. That's in theory. The theory is championed by Armando José Yáñez Soler, of Elda Alicante, Spain. The town of Elda Alicante, until now, has been best known as the home of the Museo del Calzado (the Museum of Footwear), but if Yáñez's theory is correct, his fame could eventually surpass that of the museum.

Yáñez published details of his research in the journal *Medical Hypotheses*. 'The human being has evolved to become a naked monkey', he writes, but 'there is no apparent reason to continue the evolutionary process up to becoming a bald monkey.' Common baldness 'is a degenerative process derived from certain inadequate cultural practices, such as excessive hair cutting or certain types of haircuts'.

The process is roughly akin, in his view, to the coming of Alzheimer's disease. 'It is generally accepted', he notes, citing a small study that appeared in the journal *Neurology* in 2001, 'that loss of memory in people over 60 years of age is mainly due to ... certain behaviours of the individual.'

Yáñez is fascinated by sebum, the oily secretion produced by tiny glands in the regions of skin where hair is produced. The sebum flows around and through the hair. If this gunk builds up, says Yáñez, there ensues a cascade of physiological events that lead to baldness.

Combing, brushing, touching, or massaging one's hair helps keep the sebum flowing out of the scalp. Sleeping with one's head on a good, absorbent pillow sops it up. Yáñez is almost lyrical in explaining the rise and meanderings of the sebum and the attraction of pillows.

Pillows are just the half of it. Luxuriant, flowing hair is the other half. Yáñez explains that sebum can move out and along the lengthy surface of each hair and so eventually ooze its way to a pillow or a hairbrush or some other absorbent, sebum-sucking surface. Somehow, short hair doesn't cut it.

Yáñez says his theory explains why baldness is more common in men than in women. 'Nature provides both sexes with the capacity to have long hair', he points out. And people with thick or curly hair have especially good sebum-elimination. 'This explains why certain ethnicities or cultures such as Native Americans, Rastafaris, Gypsies, etc. do not suffer from common baldness.' Furthermore, 'many people affected by common baldness have noted that they started to suffer from it during military service.... The difference in hair length is the key. Military people, skinheads and others wear their hair short and therefore they can induce problems with sebum flow.'

Yáñez says he is 'aware that this thought-provoking theory will give rise [to] a lot of skepticism'.

Yáñez Soler, Armando José (2004). 'Cultural Evolution as a Possible Triggering or Causative Factor of Common Baldness.' *Medical Hypotheses* 62 (6): 980–85.
Scarmeas, G. Levy, M.-X. Tang, J. Manly, and Y. Stern (2001). 'Influence of leisure activity on the incidence of Alzheimer's Disease.' *Neurology* 57: 2236–42.

A HOLE IN THE HEAD

Holes seem simple enough, until you examine them closely. Marco Bertamini, of the University of Liverpool, and Camilla Croucher, of the University of Cambridge, peered at one particular aspect. Their study, called 'The Shape of Holes', appeared in the journal *Cognition*. In it, they say: 'We discuss the many interesting aspects of holes as a subject of study in different disciplines, and predict that much insight especially about shape will continue to come from holes.'

'The Shape of Holes' is a more specialized report than its title implies. Its central question deals with how we see and understand edges: do the contours of a hole belong to the hole, or to the surrounding object? Psychologists, philosophers, artists, and, recently, also computer scientists have wrestled with this and with each other for the better part of a century.

Almost certainly you have played with the black-and-white drawing made famous by Danish psychologist Edgar Rubin. That's the drawing in which you can choose to see either two faces or a vase – but not both at the same time. Look at that drawing again now, paying attention to the border between black and white, and you will see the nature of this hole question.

Bertamini and Croucher had volunteers look at line drawings that include particular kinks and bends. The goal: to better understand how we use such details to perceive particular shapes. The result: Bertamini and Croucher say that, to human eyes, the edges of a hole are not themselves part of the hole.

There is a rich, deep history of people looking into holes. Everyone, it appears, is aware of the oddity of the enterprise. In 1970, the *Australasian Journal of Philosophy* published an instant classic of hole scholarship. Written by Princeton University philosopher David Kellogg Lewis and his wife, Stephanie, it bears the simple title 'Holes'. 'Holes' has been described as a 'whimsical dialogue debating the ontological nature of holes'.

Recently, Flip Phillips, J. Farley Norman, and Heather Ross explored holes by using twelve sweet potatoes. They conducted their experiment at Western Kentucky University.

This project required careful preparation. The team cast silhouettes of the sweet potatoes on to a projection screen, photographed the silhouettes with a digital camera, transferred the digitized pictures to a Macintosh computer, and then fed the data to a laser printer. The result: sheets of paper imprinted with potato silhouettes. The scientists then recruited volunteers. They asked the volunteers to copy each potato silhouette to an adjacent blank area, paying special attention to the dents and protuberances of each potato-shape. The results confirm an old theory that dents and nubs play a big part in how we recognize shapes.

In theory, this patches a gap in our understanding of holes and other shapes: at the edges, it's the kinks – not the long smooth stretches – that matter most.

Bertamini, Marco, and Camilla J. Croucher (2003). 'The Shape of Holes.' *Cognition* 87 (1): 33–54.
Lewis, David Kellogg, and Stephanie R. Lewis (1970). 'Holes.' *Australasian Journal of Philosophy* 48: 206–12; reprinted in Lewis, D. K. (1983). *Philosophical Papers*, vol. 1. New York: Oxford University Press, 3–9.
Norman, J. F., F. Phillips, and H. E. Ross (2001). 'Information Concentration Along the Boundary Contours of Naturally Shaped Solid Objects.' *Perception* 30: 1285–94.

COLOUR PREFERENCE IN THE INSANE

In the summer of 1931, Siegfried E. Katz of the New York State Psychiatric Institute and Hospital published a study in the *Journal of Ab-*

normal and Social Psychology called 'Color Preference in the Insane'. Assisted by a Dr Cheney, Katz tested 134 hospitalized mental patients, asking them about their favourite colours. For simplicity's sake, he limited the testing to six colours: red, orange, yellow, green, blue, and violet. No black. No white. No shades of grey.

COLOR PREFERENCE IN THE INSANE [*]

By S E. KATZ

NEW YORK STATE PSYCHIATRIC INSTITUTE AND HOSPITAL

PROBLEM

THE object of this study was to ascertain some of the factors which influence the preference for certain colors among persons affected with mental diseases. The writer has endeavored to determine: (1) whether certain colors are generally pleasing to the insane, (2) whether noticeable similarities exist between color preference in the insane and in children, (3)

'These colors', he wrote, 'rectangular in shape, one and one-half inches square, cut from Bradley colored papers were pasted in two rows on a gray cardboard. They were three inches apart. The colors were numbered haphazardly and the number of each color placed above it. The cardboard was presented to the patient and he was asked to place his finger on the number of the color he liked best. After he had made the choice he was asked in a similar manner for the next best color, and so on.'

Some of the patients 'cooperated well', and made six choices. Others, Katz reported, 'quickly lost interest and made only one, two or three'.

Blue was the most popular colour. Men, in the aggregate, then favoured green, but women patients were divided on green, red, or violet as a second choice.

Patients who had resided in the hospital for three or more years were slightly less emphatic about blue. Katz says that these long-term guests were 'those with most marked mental deterioration'. Their preference, as a group, shifted somewhat towards green and yellow. Those of longest tenure, though few in number, had a slightly elevated liking for orange.

The report is packed with titbits that beg, even now, for further analysis:

- Blue was given 'first preference' by thirty-eight percent of schizophrenics and manic depressives, versus forty-two percent of other patients.
- Green was first choice among sixteen percent of schizophrenics, nine percent of manic depressives, and thirteen percent of other patients.
- Red was first choice for sixteen percent of manic depressives, twelve percent of schizophrenics, and fifteen percent of other patients; it was the second choice of twenty-two percent of manic depressives, eighteen percent of schizophrenics, and thirteen percent of other patients.
- Orange and yellow gained the highest 'best liked' rating among manic depressives; green among schizophrenics, and violet among other patients.

Katz foresaw practical applications for his research. He suggested that 'in the furnishings of living quarters the selection of colors pleasing to special groups of patients might be worth consideration'.

Consciously or not, hospital staff seem to have followed Dr Katz's insights in fashioning their own, personal, at-work appearance. The evocatively named Bragard Medical Uniforms, a New York firm founded in 1933, now publishes a list of the most popular uniform colours. The list currently is topped by, in order: royal blue; dark grey (which, alas, Katz excluded from his survey); dark green; and red.

Katz, Siegfried E. (1931). 'Color Preference in the Insane.' *Journal of Abnormal and Social Psychology* 26 (2): 203–11.

MAY WE RECOMMEND

'AN EXPERIMENT IN DREAM TELEPATHY WITH "THE GRATEFUL DEAD"'

by Stan Krippner, Monte Ullman, and Bob Van de Castle (published in the *Journal of the American Society of Psychosomatic Dentistry and Medicine*, 1973)

NEUROLOGICAL DAMAGE FROM PRAYING

He who prays fervently courts danger – neurological danger.

This stark fact has only recently been reported to the public, in a study published by five neurologists at Christian-Albrechts-University in Kiel. But fear not – the risk for any particular individual is low. In the recorded history of the world, the physicians try to assure us, this is probably the very first case.

Perhaps to avoid inciting jitters among the devout, perhaps due to worries about inter-cultural tensions, or perhaps merely through professional tradition, the doctors couch their tale in dry, medico-lingo-laced language: 'We report on an unusual presentation of a task-specific focal oromandibular dystonia in a 47-year-old man of Turkish descent. His speech was affected exclusively while reciting Islamic prayers in Arabic language, which he otherwise did not speak.'

The problem – involuntary jaw-muscle twitching – had been cropping up for two years. It occurred *only* when the man recited the Arabic prayers he had been praying since childhood, never at other times. It happened whether he said the prayers loud and fast, or just muttered them slowly. It would cease right after he finished praying. It never happened when he spoke in German or in Turkish. An extensive battery of tests showed him to be otherwise in good neurological, muscular, and dental health.

The doctors came up with a simple, fairly effective fix: they had the man touch himself lightly on the jaw. Usually, this would interrupt the spooky jaw gyrations.

This general kind of problem is called 'focal dystonia'. It's the involuntary fluttering of muscles that one ordinarily controls masterfully. It arises, somewhat mysteriously, in a few extraordinarily unlucky people who perform 'a highly stereotyped and frequently repeated motor task'. It's what happens in writer's cramp, and in the eyelid twitching known as blepharospasm, and very occasionally in certain specialized professions. Doctors have seen it in pianists, tailors, and assembly-line workers. But never before in someone whose repetitive action consisted only of saying prayers.

The doctors, Tihomir Ilic, Monika Pötter, Iris Holler, Günther Deuschl, and Jens Volkmann, appear to have been surprised – and possibly a bit delighted. They bestowed upon this condition the name 'praying-induced oromandibular dystonia'.

And digging through medical histories, they did find one earlier case that seems truly analogous. It was recorded in England in the early 1990s. The doctors (N. J. Scolding of London and four colleagues) involved in the case later published an account. 'A 33-year-old right-handed agricultural auctioneer first noticed involuntary deviation of his jaw to the right developing while auctioneering', they wrote. 'Further attempts to conduct auctions met with an inevitable recurrence of his symptoms, usually after 2 to 3 minutes of speaking, and eventually it was necessary for him to move to a clerical job.'

Those doctors, like their later German counterparts, suffered the involuntary fluttering of heart and mind that is triggered by the beckoning finger of fame. They concocted a new piece of medical jargon. Thus did the term 'auctioneer's jaw' enter medical literature.

Ilic, Tihomir V., Monika Pötter, Iris Holler, Günther Deuschl, and Jens Volkmann (2005). 'Praying-Induced Oromandibular Dystonia.' *Movement Disorders* 20 (3): 385 86.
Scolding, N. J., S. M. Smith, S. Sturman, G. B. Brookes, and A. J. Lees (1995). 'Auctioneer's Jaw: A Case of Occupational Oromandibular Hemidystonia.' *Movement Disorders* 10 (4): 508v9.

TWO

THINGS THAT MATTER

AN IMPROBABLE INNOVATION

'CIRCULAR TRANSPORTATION FACILITATION DEVICE'
a/k/a 'the wheel' by John Keogh (Australian Innovation Patent no. 2001100012, granted and honoured, jointly with the patent office, with the 2001 Ig Nobel Prize in technology)

Some of what's in this chapter: Breaching, by means of a shoe • Making horrid, chilling sounds • Skipping and hopping • Plumbing the masticatory mechanism, aurally • Glug-glugging • OmmmmmmmmOMmmmmmOMmm • Self-encrustation with bees and music • Foreseeing football on Mars • Basic black, in the desert • Walking with washing machines • A strapless dress, forcefully

THE GREAT UNTIED-SHOELACE EXPERIMENT

Details about the late Norbert Elias's international untied-shoelace experiments were difficult to track down. But Ingo Mörth found them.

Mörth, a professor at the Johannes Kepler Universität in Linz, Austria, broke the news in an article called 'The Shoe-lace Breaching Experiment', published in the June 2007 issue of *Figurations: Newsletter of the Norbert Elias Foundation*: 'Norbert Elias started a series of breaching experiments, beginning ad hoc, and ending in various situations in Spain, France, England, Germany, and Switzerland. He strolled around in all these contexts with intentionally untied and trailing shoe-laces.'

Elias had an eminent career as a sociologist, beginning in Germany in the 1930s. After retiring as a reader at University of Leicester in 1964, he went a-wandering, doing sociological research as a byproduct of his tourism.

In the Spanish fishing village of Toremolinos in 1965, giggling girls spurred him to realize that his left shoelaces 'were untied and trailing'. Mörth describes the magic that resulted: 'By retying the loose shoe-lace, Elias had the feeling of being included in the village community – at least for a moment, and based on the community aspect of the everyday reality in the village, people took notice and nodded approval of his rectifying something that had a disturbing appearance.'

Elias thereupon began his experiments, strolling across Europe with deliberately untied shoelaces. In England 'mostly elderly gentlemen reacted by communicating with me on the danger of stumbling and falling'. In Germany 'older men only looked at me somewhat contemptuously, whereas women reacted directly and tried to 'clean up' the obvious disorder, on the tram as well as elsewhere'.

The professor and his laces thusly pioneered – though the academic world mostly failed to celebrate him and them for it – what are now known as 'breaching experiments'. American sociologist Harold Garfinkel coined the term and rose to fame by conducting a series of such activities. As Mörth explains, these experiments 'breached the taken-for-granted assumptions underlying everyday situations, thereby generating consternation and embarrassment among other people present'.

Elias's many fans in the Norbert Elias Foundation and elsewhere were aware that he had done something with shoelaces, but because Elias did not publish a formal academic study, most did not know that they could read a firsthand account of what, where, and when he did it. Thanks to Mörth, scholars can now learn that Elias's historic report was published in the German weekly magazine *Die Zeit* in 1967, in the travel section, under the headline 'Die Geschichte mit den Schuhbändern' ('The Story of the Shoe-Laces').

In publicizing the existence of Elias's original account, Mörth flung open a door through which researchers had, for forty years, thought themselves restricted to only a squinting glance.

Mörth, Ingo (2007). 'The Shoe-lace Breaching Experiment.' *Figurations: Newsletter of the Norbert Elias Foundation* 2 (27): 4–6.
Elias, Norbert (1967). 'Die Geschichte mit den Schuhbändern.' *Die Zeit*, 17 November.

AN IMPROBABLE INNOVATION

'GARMENT DEVICE CONVERTIBLE TO ONE OR MORE FACEMASKS'
a/k/a a brassiere that, in an emergency, can be quickly converted
into a pair of protective masks, by Elena N. Bodnar, Raphael C. Lee,
and Sandra Marijan (US Patent no. 7,255,627, granted 2007 and
honoured with the 2009 Ig Nobel Prize in public health)

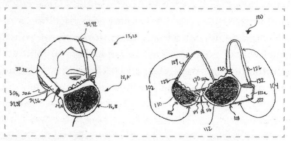

From 'Garment Device Convertible to One or More Facemasks'

CHILLING SOUNDS

Fingernails on a blackboard. Why does the very phrase send chills
down one's back? The question has annoyed scientists for at least
2300 years. Aristotle mentioned the existence of what he called 'hard
sounds', but didn't try very hard to explain them.

In the mid-1980s, three scientists assaulted the problem directly,
subjecting volunteers to a battery of electronically synthesized nails-
on-blackboard screeching. D. Lynn Halpern, Randolph Blake, and
James Hillenbrand at Northwestern University in Evanston, Illinois,
published details in the journal *Perception and Psychophysics*. They called
their study 'Psychoacoustics of a Chilling Sound'.

First, they ran some tests to establish exactly where nails-on-a-
blackboard ranks in the hierarchy of annoying sounds.

They recruited a panel of volunteers – a different group from the
one that would later undergo the intensive, dedicated exposure to the
Sound of Sounds. The panel listened to recordings of sixteen different,
hypothetically 'annoying' sounds. They rated just how annoying each
was. This ranged from not very (for chimes, rotating bicycle tyres, and

running water) to excruciating. Jingling keys mildly annoyed some people. Then, increasingly less pleasant, came the sounds of a pencil sharpener; a blender motor; a dragged stool; a metal drawer being opened; scraping wood; scraping metal; and rubbing two pieces of Styrofoam together. But the annoyance of fingernails-on-a-blackboard topped them all.

Halpern, Blake, and Hillenbrand, having established this simple fact, then converted the tape recording to a digital signal, so that they could manipulate and experiment with constituent high and low pitches. The formal report notes the researchers' belief that the signal was of good quality. 'To the authors and several other reluctant volunteer listeners', they write, 'the digitized, filtered signal sounded very similar to, and just as unpleasant as the original.'

The original recorded, pre-digitized sound was not actually of scraping fingernails, but of something known, from previous experiments, to be very like it. In a footnote, Halpern, Blake, and Hillenbrand confide that 'the instruction set used in this study included a description of [a] three pronged garden tool being dragged across a slate surface. Virtually all subjects shuddered upon reading this portion of the instructions.'

The shuddering volunteers listened to several different, digitally doctored versions of the sound, and rated the unpleasantness of each.

The study's conclusion, when all was scraped and done, is perhaps worth quoting: 'Our results demonstrate that the unpleasant quality associated with the sound of a solid object scraped across a chalkboard is signaled by acoustic energy in the middle range of frequencies audible to humans. High frequencies, contrary to intuition, are neither necessary nor sufficient to elicit this unpleasant association. Still unanswered, however, is the question of *why* this and related sounds are so grating to the ear.'

The story didn't end there, of course. In 2004, Josh McDermott and Marc Hauser of Harvard University conducted a series of acoustic and psychological experiments. In the process, they discovered a major difference between Harvard students and cotton-top tamarin monkeys. Harvard students actively avoid fingernail-on-blackboard sounds when given the opportunity, but tamarins generally don't. McDermott and Hauser hazard some guesses about why this is so –

they suggest it may be in some way related to our ability to appreciate or deplore music.

The 2006 Ig Nobel Prize in the field of acoustics was awarded to Halpern, Blake, and Hillenbrand for their chilling research. Still, the mystery endures, giving cold discomfort to almost everyone who hears about it.

Halpern, D. Lynn, Randolph Blake, and James Hillenbrand (1986). 'Psychoacoustics of a Chilling Sound.' *Perception and Psychophysics* 39: 77–80.
McDermott, Josh, and Marc Hauser (2004). 'Are Consonant Intervals Music to Their Ears? Spontaneous Acoustic Preferences in a Nonhuman Primate.' *Cognition* 94: B11–B21.

HOP, SKIP, AND REACH CONCLUSIONS

When do young adults skip and hop, and why? These are the questions raised by Allen W. Burton, Luis Garcia, and Clersida Garcia. Their answers appear in the research report 'Skipping and Hopping of Undergraduates: Recollections of When and Why'. Burton, at the University of Minnesota, and Garcia and Garcia, at Northern Illinois University, write that 'the purpose of this study was to compare the reasons why young adults skip and hop and when they last skipped and hopped'.

The researchers collected data from 253 female and 411 male university students. Each of those young adults was asked two skipping and two hopping questions:

1) Approximately how long ago did you last spontaneously skip?
2) Why did you skip? In other words, what elicited your skipping behaviour?
3) Approximately how long ago did you last spontaneously hop?
4) Why did you hop? In other words, what elicited your hopping behaviour?

Based on the results of that survey, Burton, Garcia, and Garcia conclude that hopping and skipping are not the same thing. Not to undergraduate students. At least, not as far as when and why are concerned. At least, not completely. Their report explains in detail.

Top Four Responses of Women and Men to the Question: 'When Did You Last Spontaneously Skip/Hop?' Plus Total of 24 Affect Responses

Why?	Skip		Hop	
	Women n: 266	Men n: 426	Women n: 266	Men n: 426
Don't know, no reason, no idea, no clue	3.4	12.9	8.3	19.9
Never, don't remember when, skipped question	3.0	4.5	9.4	4.1
Drills, physical education drills, had to (generally not spontaneous)	2.3	5.9	2.6	5.0
General play (not 'playing around'), simple games (e.g., tag, sack race)	6.4	4.5	6.8	6.2
Go over obstacle	0.4		1.5	5.5
Hopscotch, hop in squares/markings/patterns on surface	0.4		9.4	5.0
While playing specific sport or performing specific skill	0.8		0.8	7.4
Funny, silly, kidding, joking, laughing, humour, slap-happy, make fun of, make laugh, embarrass	6.4		1.1	1.0
Happy, light-hearted, good mood, playful, joyful, giddy	18.0		3.8	2.6
Stupid, dumb, crazy, dorky, strange, weird, goofing, messing, fooling, geeking, playing, screwing around	11.3		3.0	2.4
Total affect (24 categories)	57.1	51.3	24.1	19.2

That's the story on the when and why of hopping and skipping. Now, how about the how?

Claire Farley, Reinhard Blickhan, Jacqueline Saito, and Richard Taylor at Harvard University published a massive six-page report on their experiments with 'hopping frequency in humans'. Two young women and two young men did the hopping, individually, on a treadmill. The treadmill ran, so to speak, at various speeds. Each individual turned out to have a preferred hopping frequency, at which she or he most strongly resembled (in certain respects) a rock glued atop a spring.

That's true of hopping on two feet. Hopping on one foot is an entirely different question. Or at least it has the potential to be an entirely different question. That potential was explored in research done by G. P. Austin, G. E. Garrett, and D. Tiberio at Sacred Heart University, in Fairfield, Connecticut. In June 2002 they leaped into public view with a report entitled 'Effect of Added Mass on Human Unipedal Hopping'. Six months later they popped up again, with 'Effect of Frequency on Human Unipedal Hopping'. The next year they jumped into sight yet again, with 'Effect of Added Mass on Human Unipedal Hopping at Three Frequencies'.

Have they gained a leg up on their professional competitors? How high will their ambitious research programme take them? We shall see.

Burton, Allen W., Luis Garcia, and Clersida Garcia (1999). 'Skipping and Hopping of Under-graduates: Recollections of When and Why.' *Perceptual and Motor Skills* 88: 401–6.
Farley, Claire T., Reinhard Blickhan, Jacqueline Saito, and C. Richard Taylor (1991). 'Hopping Frequency in Humans: A Test of How Springs Set Stride Frequency in Bouncing Gaits.' *Journal of Applied Physiology* 71 (6): 2127–32.
Austin, G. P., G. E. Garrett, and D. Tiberio (2002). 'Effect of Added Mass on Human Unipedal Hopping.' *Perceptual and Motor Skills* 94 (3): 834–40.
Austin, G. P., D. Tiberio, and G. E. Garrett (2002). 'Effect of Frequency on Human Unipedal Hopping.' *Perceptual and Motor Skills* 95 (3): 733–40.
Austin, G. P., D. Tiberio, and G. E. Garrett (2003). 'Effect of Added Mass on Human Unipedal Hopping at Three Frequencies.' *Perceptual and Motor Skills* 97 (2): 605–12.

SOUNDS DELICIOUS

Can a machine identify what you're chewing, merely from the sound? Yes, if you are at a laboratory in Zurich, Switzerland, or Hall-in-Tirol, Austria, and if you are chewing potato crisps, apples, mixed lettuce, cooked pasta, or boiled rice.

Oliver Amft, Mathias Stäger, and Gerhard Tröster of the Swiss Federal Institute of Technology, and Paul Lukowicz of Austria's University for the Health Sciences, Medical Informatics and Technology (UMIT), describe their work succinctly: 'using wearable microphones to detect and classify chewing sounds (called mastication sounds) from the user's mouth'. But, they explain, this is just stage 1 of their dream. It's an unusual dream: to build a computer-based machine 'that precisely and 100% reliably determines the type and amount of all and any food that the user has consumed'.

Nothing about stage 1 is easy. The scientists list three different approaches that a machine might take in trying to sense someone's food intake automatically:

a)   detecting and analysing chewing sounds,

b)   using electrodes mounted on the base of the neck (e.g., in a collar) to detect and analyse bolus swallowing,

c)   using motion sensors on hands to detect food intake-related motions.

Amft, Stäger, Tröster, and Lukowicz chose option (a). It, alone, seemed within the range of the technology available to them today.

Their report is written for specialists, but contains delights for everyone. My favourite is the graph titled 'Chewing sound and speech recording in a room with background music', which depicts the sound intensity during a minute-long span. The graph's four segments are labelled 'eating lettuce', 'user speaking', 'eating pasta', and 'music playing'.

Here are some of the things the scientists say they learned in having their machine analyse a total of 650 'chewing sequences' produced by four healthy chewers:

•   Good quality chewing sound signal can be obtained by placing a microphone in the ear canal.

•   Chewing sounds can be discriminated from a signal containing a mixture of speech, silence and chewing.

•   Listening to a sequence of chewing sounds, it is possible to identify the beginnings of the individual single chews.

•   Chewing-sound-based discrimination between very different kinds of food – the kinds mentioned above – is possible with greater than eighty percent accuracy.

This all builds on decades of work that began with Swedish Institute for Food Preservation Research scientist B. K. Drake's 1963 study 'Food Crushing Sounds: An Introductory Study'.

The study of chewing sounds is a very specialized field. (For an extreme example, see 'Crisp Sounds' on page 138.) The field apparently acquired a name in 1966, when British dentist D. M. Watt published a paper called 'Gnathosonics: A Study of Sounds Produced by the Masticatory Mechanism'.

Amft, Stäger, Tröster, and Lukowicz are proud of their chew-sound-analysis achievement. But mindful of technology's limits, they aim to keep their aims simple. In their words: 'The system does not need be fully automated to be useful ... it is perfectly sufficient if at the end of the day the system can remind the user that for example "at lunch you had something wet and crisp (could have been salad) and some soft texture stuff (spaghetti or potatoes)" and asks him to fill in the details.'

Amft, Oliver, Mathias Stäger, Paul Lukowicz, and Gerhard Tröster (2005). 'Analysis of Chewing Sounds for Dietary Monitoring.' *UbiComp 2005: Proceedings of the 7th International Conference on Ubiquitous Computing*, Tokyo, Japan, 11–14 September: 56–72.
Drake, B. K. (1963). 'Food Crushing Sounds: An Introductory Study.' *Journal of Food Science* 28 (2): 233–41.
Watt, D. M. (1966). 'Gnathosonics: A Study of Sounds Produced by the Masticatory Mechanism.' *Journal of Prosthetic Dentistry* 16 (1): 73–82.

POUR LAWS

On the glug-glug of ideal bottles

By CHRISTOPHE CLANET AND GEOFFREY SEARBY
Institut de Recherche sur les Phénomènes Hors Equilibre,

We present an experimental study of the emptying of an ideal vertical bottle under gravity g. The idealization reduces the bottle to a cylinder of diameter D_0, length L, closed at the top and open at the bottom through a circular thin-walled hole of diameter d, on the axis of the cylinder. The study is performed in the low-viscosity limit. The oscillatory emptying of the 'bottle' is referred to as the glug-glug, and is

When physics professors take to the bottle, they can be tenacious about it. Take Christophe Clanet and Geoffrey Searby, who wrote a highly condensed, fourteen-page report called 'On the Glug-Glug of Ideal Bottles', which was published in the *Journal of Fluid Mechanics*. Like so much of the literature emerging from Europe during the past two centuries, this study celebrates what happens when liquid is poured from a container.

The pair of de-bottling experts is based at the Institut de Recherche sur les Phénomènes Hors Equilibre in Marseille, France. Both men are fascinated by bubbles and motion. Searby heads a French–German committee doing research on rocket engine combustion, while Clanet has become tops in the physics subspecialty of skipping stones across ponds.

'Glug glug' is now a technical term, thanks mostly to Clanet and Searby. They tried it out at a physics conference in 1997, presenting a talk called, plainly, 'On the Glug-Glug of the Bottle'. Their opening words were circumspect: 'We study experimentally the emptying of a vertical cylinder of diameter D and length L.' The audience response was such that Clanet and Searby continued their exploration of glug-glugs. They delved into the theoretical aspects, as well as the empirical.

Their follow-up paper begins with a dramatic sentence: 'An image of life is a return to the thermodynamic equilibrium of death via the oscillations of our heartbeats.' Then, with a quick literary pirouette, they describe the 'onomatopoeic glug-glug' of an emptying bottle. 'This oscillatory behaviour', they remind us, 'starts at the opening and continues until the bottle is empty.'

The apparent weight of the bottle lurches way up and slightly less down, up, down, up, down, until the liquid is gone. Clanet and Searby produced a graph of this behaviour, a visual form of glug-glug that some scientists find as delightful as the sound.

The experiment involved Newtonian liquid, a tank, two valves, a pump, a pressure sensor, a camera, and a laser beam. It built upon the pioneering bubble-behaviour work done in the late 1940s by Geoffrey Taylor at the University of Cambridge. Taylor's bubbles inspired a ragged, international line of experimentation on bottle-emptying that culminated with the Clanet/Searby glug-glug work.

The fruits of the experiment are sweet. Through painstaking work, Clanet and Searby elucidated the basic law of glug-glug: the time needed to empty a bottle depends on the diameter of the bottle, and also on the diameter of the hole.

Of course, this is the law for an idealized bottle shape – a can rather than the beloved Coca-Cola bottle or other quirky form. Even for a Coke can, though, there remains the open question of the tab-shaped opening. Clanet and Searby used a cylinder with a circular hole. Whether and how much a different hole shape affects the glug-glug is, almost needless to say, a matter for further research.

Clanet, Christophe, and Geoffrey Searby (2004). 'On the Glug-Glug of Ideal Bottles.' *Journal of Fluid Mechanics* 510: 145–68.

Clanet, C., G. Searby, and E. Villermaux (1997). 'On the Glug-Glug of the Bottle.' American Physical Society, Division of Fluid Dynamics Meeting, 23–25 November, abstract #Df.10.

Davies, R. M., and Geoffrey Taylor (1950). 'The Mechanics of Large Bubbles Rising Through Liquids and Through Liquids in Tubes.' *Proceedings of the Royal Society of London, Series A* 200 (22 Feb): 375–90.

THE REPETITIVE PHYSICS OF OM

Two Indian scientists are wielding sophisticated mathematics to dissect and analyse the traditional meditation chanting sound 'Om'. The Om team has published six monographs in academic journals. These plumb certain acoustic subtleties of Om, which the researchers say is 'the divine sound'.

Om has many variations. In a study published in the *International Journal of Computer Science and Network Security*, the researchers explain: 'It may be very fast, several cycles per second. Or it may be slower, several seconds for each cycling of [the] Om Mantra. Or it might become extremely slow; with the mmmmmm ... sound continuing in the mind for much longer periods, but still pulsing at that slow rate. It is somewhat like one of these vibrations:

OMmmOMmmOMmm ...
OMmmmmOMmmmmOMmmmm ...
OMmmmmmmmmOMmmmmmmmmOMmm

The important technical fact is that no matter what form of Om one chants at whatever speed, there is always a basic Omness to it.

Ajay Anil Gurjar and Siddharth A. Ladhake published their first Om paper, titled 'Time-Frequency Analysis of Chanting Sanskrit Divine Sound "OM"', in 2008. Ladhake is the principal at Sipna's College of Engineering and Technology in Amravati, India. Gurjar is an assistant professor in that institution's department of electronics and telecommunication. Both specialize in electronic signal processing. They now subspecialize in analysing the one very special signal.

In their introductory paper, Gurjar and Ladhake explain (in case there is someone unaware of the basics) that: 'Om is a spiritual mantra, outstanding to fetch peace and calm. The entire psychological pressure and worldly thoughts are taken away by the chanting of Om mantra.'

Analysis Of Acoustic of 'OM' Chant To Study It's Effect on Nervous System

Ajay Anil Gurjar , Siddharth A. Ladhake, Ajay P. Thakare
Sipna's College of Engineering & Technology, Amravati (Maharashtra), India

Summary

OM does not have a translation. Therefore, the Hindus consider it as the very name of the Absolute, it is body of sound. In the scriptures of ancient India, the OM is considered as the most powerful of all the mantras. The others are considered aspects of the OM, and the OM is the matrix of all other mantras. It has been recognized that the Mantras have beneficial effects on the human being. It is one phrase. The syllable OM is quite familiar to a Hindu. It occurs in every prayer. Invocation to most gods begins with this syllable. OM is also pronounced as AUM. The syllable OM is not specific to Indian culture. It has religious significance in other religions also. Although OM is not given any specific definition and is considered to be a ...

personally in tune. The use of this mantra can be profound. At first, it is best to use the mantra gently and for short periods of time. The insights from the OM mantra can be significant, and it is good to integrate the insights gradually with daily life.

2. Om Mantra and Methods of Practice

It is proposed by Swami Jnaneshvara Bharti that there are many rhythms in the body and mind, both gross and subtle. The sound of OM, rising and falling, at whatever ...

No one has explained the biophysical processes that underlie this fetching of calm and taking away of thoughts. Gurjar and Ladhake's time-frequency analysis is a tiny step along that hitherto little-taken branch of the path of enlightenment.

They apply a mathematical tool called wavelet transforms to a digital recording of a person chanting 'Om'. Even people with no mathematical background can appreciate, on some level, one of the blue-on-white graphs included in the monograph. This graph, the authors say, 'depicts the chanting of "Om" by normal person after some days of chanting'. The image looks like a pile of nearly identical, slightly lopsided pancakes held together with a skewer, the whole stack lying sideways on a table. To behold it is to see, if nothing else, repetition.

At the end, Gurjar and Ladhake write: 'Our attentiveness and our concentration are pilfered from us by the proceedings take place around us in the world in recent times ... By this analysis we could conclude steadiness in the mind is achieved by chanting OM, hence proves the mind is calm and peace to the human subject.'

Much as people chant the sound 'Om' over and over again, Gurjar and Ladhake repeat much of the same analysis in their other five studies, managing each time to chip away at some slightly different mathematico-acoustical fine point.

Gurjar, Ajay Anil, and Siddharth A. Ladhake (2008). 'Time-Frequency Analysis of Chanting Sanskrit Divine Sound "OM"'. *International Journal of Computer Science and Network Security* 8 (8): 170–75.
— (2009). 'Spectral Analysis of Sanskrit Divine Sound OM.' *Information Technology Journal* 8: 781–85.
— (2009). 'Optimal Wavelet Selection For Analyzing Sanskrit Divine Sound "OM"'. *International Journal of Mathematical Sciences and Engineering Applications* 3 (2). 225–33.
— (2009). 'Analysis of Speech Under Stress Before and After *OM* Chant Using MATLAB 7.' *International Journal of Emerging Technologies and Applications in Engineering, Technology and Sciences* 2 (2): 713–18.

— (2009). 'Time-Domain Analysis of "OM" Mantra to Study It's [sic] Effect on Nervous System.' *International Journal of Engineering Research and Industrial Applications* 2 (3): 233–42.

Gurjar, Ajay Anil (2009). 'Multi-Resolution Analysis of Divine Sound "OM" Using Discrete Wavelet Transform.' *International Journal of Emerging Technologies and Applications in Engineering, Technology and Sciences* 2 (2): 468–72.

Gurjar, Ajay Anil, Siddharth A. Ladhake, Ajay P. Thakare (2009). 'Analysis of Acoustic [sic] of "OM" Chant to Study It's [sic] Effect on Nervous System.' *International Journal of Computer Science and Network Security* 9 (1): 363–67.

HUMMING IN THE KEY OF BEE

Norman E. Gary is the rare academic who plays clarinet while he is covered with live bees, and often does so in public.

An emeritus professor of apiculture at the University of California (Davis), Gary also plays Dixieland music in a human ensemble called the Beez Kneez Jazz Band. He generally goes solo – he alone with his instrument – for the bee-encrusted gigs.

Hollywood has used Gary's bee-wrangling talents and sometimes his acting ability, though seldom his clarinet, in more than a dozen films. Among them: *The X Files, Fried Green Tomatoes, Invasion of the Bee Girls,* and *Candyman: Farewell to the Flesh.*

Several of Gary's scientific activities involve vibration, a general physics phenomenon of which music is just a part. He has microwaved bees. He has also analysed one of the lesser-known (to most humans) sounds that bees produce. Details appear in a 1984 monograph published with colleague S. S. Schneider in the *Journal of Apicultural Research.* They gave their article the title '"Quacking": A Sound Produced by Worker Honeybees after Exposure to Carbon Dioxide'.

Gary has vacuumed bees. He has also made it easier and more efficient for others who want or need to vacuum the insects, by inventing a purpose-built bee vacuum with his colleague Kenneth Lorenzen. The wording in their patent could, with a bit of work, be set to hummable music: 'By the operation of the mechanism in the fashion disclosed, the bees on the opposite sides of a comb, and eventually of a plurality of combs and frames, are removed therefrom by a concomitant vacuuming and brushing operation.'

The professor has published more than one hundred academic papers, many of them about bees. In one of the earliest, called 'The Case of Utter vs. Utter', he took a fond look back at a court case decided in 1901 in Goshen, New York, starring two brothers from a family named Utter.

The brothers disagreed – Utterly, of course – about many things. The question here was: did the bees associated with one brother, a bee-keeper, eat the peaches growing on trees owned by the other brother, a fruit grower? Perhaps the most enjoyable account appeared soon after the trial, in the *Rocky Mountain Bee Journal*. The anonymous writer says: 'It was amusing to see the plaintiff try to mimic the bee, on the witness stand as he swayed his head from one side to the other, raised up on his legs and flopped his arms. His motions were so utterly ridiculous and so contrary to the real acts and achievements of the bees, that everyone in the courtroom, including the jury, laughed, and laughed heartily.'

The court ruled against that Utter, and for the other. This established a legal precedent favourable to wandering bees. It also inspired, almost sixty years later, the young Norman Gary as he began his more-than-sixty-year-long career of collaborating with and studying tiny, honey-making musicians.

Schneider. S. S., and Norman E. Gary (1984). '"Quacking". A Sound Produced by Worker Honeybees After Exposure to Carbon Dioxide.' *Journal of Apicultural Research* 23 (1): 25–30.
Gary, Norman E., and Kenneth Lorenzen (1981). 'Bee Vacuum Device and Method of Handling Bees.' US Patent no. 4,288,880.
Gary, Norman E. (1959). 'The Case of Utter vs. Utter.' *Gleanings in Bee Culture* 87 (6): 336–37.
N. A. (1901).'Bees in Court: History of the Celebrated Case of Peach Utter versus Bee-Keeper Utter.' *Rocky Mountain Bee Journal* 1 (1): 6.

VACUUM TRAVEL

The journey between London and Edinburgh would be much quicker had the London and Edinburgh Vacuum Tunnel Company been allowed and able to build a breathtaking new piece of technology, back when land was cheap and all things seemed possible. The 29 January 1825 issue of the *Mechanics Register* presents the scheme in detail: 'The London and Edinburgh Vacuum Tunnel Company is proposed to be established, with a capital of Twenty Millions Sterling, divided into 200,000 shares, of £100 each, for the purpose of forming a Tunnel or Tube of metal between Edinburgh and London, to convey Goods and Passengers between these cities and the other towns through which it passes.'

The plan is simple. There are two long tunnels or tubes, side by side, one reserved for trips northbound to Edinburgh, the other for

London-bound traffic. Boilers, located every two miles along the approximately 390-mile length of the tunnel or tube, supply steam that, through a clever bit of engineering, creates a vacuum.

At departure time, the vacuum seal is broken at the departure end, right behind the train. Thanks to the difference in pressure, the train is thus immediately impelled into the tunnel or tube.

To maintain pressure all through the journey, to keep a tight seal behind the train, there's a 'very strong air-tight sliding door, running on several small cylindrical rollers, to lessen the friction'. The inrushing air pushes the slick-sliding door. That whizzing, roller-riding door pushes the amassed railway cars onwards, onwards, faster and faster into the airless tunnel or tube.

These cars carry only freight. People never enter the tube, which, being four feet tall, is too short for most of them.

Passengers instead ride in traditional railway carriages on tracks affixed on top of the tunnel or tube. These passenger cars are coupled by strong magnets to the freight-carrying cars. As the freight train zooms through the tunnel or tube, its magnetic field drags the passenger train along on what is sure to be a rapid and exciting ride. The acceleration is such that the train travels 'altogether, in the first five minutes' of its journey, '480 miles 4448 feet'.

This would have been a considerable advance over the standard railway capabilities of the time. A dispatch in the same issue of the *Mechanics Register* allows that 'the practicality of [conventional] steam carriages for the conveyance of passengers is fully established, and we have as little doubt that the conveyance of goods at the rate of seven or eight miles an hour, will soon be as easily accomplished'.

The London and Edinburgh Vacuum Tunnel Company report is accompanied by a small notice: 'The foregoing *Jeu d'Esprit* appeared in a recent number of the *Edinburgh Star*, and being well calculated to throw ridicule upon some of the preposterous plans now before the public for the investment of money, we insert it in the *Register*.'

Nonetheless, in subsequent decades, engineers in Ireland, America, and Britain did build short stretches of pneumatic passenger railway. None spanned great distances or lasted more than a few years. Isambard Kingdom Brunel, designer of England's first great railways (and of London's Paddington station), built about twenty miles of pneumatic railway between Exeter and Newton before abandoning it as impractical.

N. A. (1825). 'London and Edinburgh Vacuum Tunnel Company, Capital 90,000 Sterling.' *Mechanics Register* 1 (13): 205–7.

MAY WE RECOMMEND

'EFFECTS OF HORIZONTAL WHOLE-BODY VIBRATION ON READING'
by Michael J. Griffin and R. A. Hayward (published in *Applied Ergonomics*, 1994)

VERY SPECIAL TOPICS

The global nature of football (a/k/a soccer in the US) varies measurably from city to city because of down-to-Earth differences in the air pressures, temperatures, and other physical conditions. But those differences are slight in comparison to the ones described in a University of Leicester study called 'Association Football on Mars'.

Calum James Meredith, David Boulderstone, and Simon Clapton published the analysis in the university's *Journal of Physics Special Topics*, which takes up topics that seldom find their way into the better-known physics journals. The journal is produced by and for university students, which makes it a bit unusual. Its unusualness quotient increases with the knowledge that the current head of the department of physics and astronomy at the University of Leicester is Professor Lester.

'Association Football on Mars' methodically calculates the altered basics of play on the red planet. 'It would be possible to retain the game in a familiar but slightly changed form', the authors say.

On the Martian surface, the gravitational pull and the air pressure are less than we're used to. The ball would encounter substantially less drag in its journeying from foot to foot to head to foot to goal. On many a kick, the ball would travel about four times as far as it would on Earth. These impressive distances come with a straightforward cost: 'the inability to "bend" the ball due to a lack of air resistance would seem to decrease the skill involved in football'.

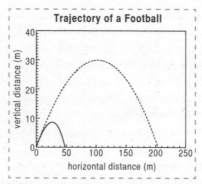

Height vs. distance for footballs kicked on Earth (solid line) and Mars (dashed line)

In the same issue of the journal, one finds other monographs by the team of Meredith, Boulderstone, and Clapton. Two of those consider a solution to our era's most pressing environmental problem.

In 'None Like It Hot', the trio propose and describe a method 'to help combat global warming by moving the Earth further [sic] away from the Sun to reduce its surface temperature'. A companion paper, 'None Like It Hot II', investigates whether this feat 'would be plausible given conventional rocket technology'. They conclude that the mass of fuel needed to perform the manoeuvre 'is only a few orders of magnitude smaller than the mass of the Earth. The number of rockets will make only a small difference due to the nature of the relationship between the two values.'

Meredith, Calum James, David Boulderstone, and Simon Clapton (2011). 'Association Football on Mars.' *Journal of Physics Special Topics* 9 (1).
— (2010). 'None Like It Hot.' *Journal of Physics Special Topics* 9 (1).
— (2010). 'None Like It Hot II.' *Journal of Physics Special Topics* 9 (1).

BASIC BLACK DRESS: HOT OR NOT?

Why do Bedouins wear black robes in hot deserts? The question so intrigued four scientists – all non-Bedouins – that they ran an experiment. Their study, called 'Why Do Bedouins Wear Black Robes in Hot Deserts?', was published in the journal *Nature* in 1980.

'It seems likely', the scientists wrote, 'that the present inhabitants of the Sinai, the Bedouins, would have optimised their solutions for desert survival during their long tenure in this desert. Yet, one may have doubts on first encountering Bedouins wearing black robes and herding black goats. We have therefore investigated whether black robes help the Bedouins to minimise solar heat loads in a hot desert.'

The research team – C. Richard Taylor and Virginia Finch of Harvard University, and Amiram Shkolnik and Arieh Borut of Tel Aviv University – quickly discovered that, as you might suspect, a black robe does convey more heat inward than a white robe does. But they doubted that this was the whole story.

They found inspiration and guidance in a 1969 report about cattle. John Hutchinson and Graham Brown of the Ian Clunies Ross Animal Research Laboratory, in Prospect, New South Wales, Australia, worked with Friesian dairy cows. The Australian team discovered that light and heat penetrate deeper into white cattle hair than into black. The saving grace for cattle is that even a tiny amount of wind whisks away that extra heat.

However, cattle are not people. So, what of man (and woman)?

Taylor, Finch, Shkolnik, and Borut measured the overall heat gain and loss suffered by a brave volunteer. They described the volunteer as 'a man standing facing the Sun in the desert at mid-day while he wore: (1) a black Bedouin robe; (2) a similar robe that was white; (3) a tan army uniform; and (4) shorts (that is, he was semi-nude)'.

Each of the test sessions (black-robed, white-robed, uniformed, and half-naked) lasted thirty minutes. It was hot there in the Negev Desert at the bottom of the rift valley between the Dead Sea and the Gulf of Elat. The volunteer stood in temperatures that ranged from a just-semi-sultry 35 degrees Celsius (95 degrees Fahrenheit) to a character-building 46 degrees Celsius (115 degrees Fahrenheit). Though he is now nameless, this was his day in the sun.

The results were clear. As the report puts it: 'The amount of heat gained by a Bedouin exposed to the hot desert is the same whether he wears a black or a white robe. The additional heat absorbed by the black robe was lost before it reached the skin.'

Bedouins' robes, the scientists noted, are worn loose. Inside, the cooling happens by convection – either through a bellows action, as the robes flow in the wind, or by a chimney sort of effect, as air rises between robe and skin.

Thus it was conclusively demonstrated that, at least for Bedouin robes, black is as cool as any other colour.

Shkolnik, Amiram, C. Richard Taylor, Virginia Finch, and Arieh Borut (1980). 'Why Do Bedouins Wear Black Robes in Hot Deserts?' *Nature* 283: 373–75.

Hutchinson, John C. D., and Graham D. Brown (1969). 'Penetrance of Cattle Coats by Radiation.' *Journal of Applied Physiology* 26 (4): 454–64.

CAPACITY OF THE NOSE

'What is the Air-Conditioning Capacity of the Human Nose?' Spring this question the next time you find yourself at a party where everybody else is an HVAC engineer. HVAC engineers specialize in heating, ventilating, and air conditioning. But, as a group, HVAC engineers are surprisingly ignorant about the air-conditioning capacity of their own noses.

Your question might throw the engineers into a two-part frenzy: first measuring each other's nasal cavity dimensions, temperatures, and vapour concentrations; and then competitively calculating, calculating, calculating until the party ends.

You could save them the trouble. Tell them about a report called 'The Air-Conditioning Capacity of the Human Nose', which was published in the *Annals of Biomedical Engineering*. There, Sara Naftali and her colleagues at Tel Aviv University tell how they attacked the question by using three artificial noses.

None of these artificial noses are ones that a mother would love if she saw one installed on her child. The first, which the scientists call 'nose-like', would seem anything but if it were mounted on someone's face. This rough-hewn product of a machine shop has internal ductwork that corresponds to 'averaged data of human nasal cavities'. A later version is called, unappealingly, 'nose-like with valve'.

The third artificial nose is a mechanically detailed reproduction of one individual's nose, with lots of twisty, bumpy idiosyncrasies. Because this nose – like most noses – is far from average, the scientists used it mostly in a sort of 'reality check' to compare against the performance of the nose-like nose and the nose-like with valve.

The ensuing artificial huffing and puffing taught them two things. First, that the nose-like noses behave realistically enough for scientists not to have to do too many uncomfortable experiments using actual people's actual noses. And second, that the basic ductwork appears to handle ninety percent or so of a person's air-conditioning needs – it delivers air of usable temperature and humidity to the lungs no matter how cold, hot, humid, or dry the atmosphere happens to be.

J. exp. Biol. 194, 329–339 (1994)   329
Printed in Great Britain © The Company of Biologists Limited 1994

THE COOLING POWER OF THE PIGEON HEAD

ROBERT ST-LAURENT AND JACQUES LAROCHELLE*
Département de biologie, Université Laval, Québec, Canada G1K 7P4

Now, should you happen to be introduced to one of the very few party-going HVAC engineers who does know the air-conditioning capacity of the human nose, do not despair. You can still stimulate a good conversation. Simply ask: what is the cooling power of the pigeon head?

For years, birders disagreed as to how their favourite animals manage to keep from overheating. More than a decade ago, Robert St. Laurent and Jacques Larochelle of the Université Laval in Quebec, Canada, wrote 'The Cooling Power of the Pigeon Head'. It describes how they inserted electronic temperature probes, via the rear exhaust openings, up into the intestines of several birds; then body-wrapped the birds; then put them into a wind tunnel.

They discovered that simply opening one's beak, without making a sound, is sufficient to keep things from getting overheated. It remains for others to see if this applies to partygoers in conversation, as well as to birds in flight.

Naftali, Sara, Moshe Rosenfeld, Michael Wolf, and David Elad (2005). 'The Air-Conditioning Capacity of the Human Nose.' *Annals of Biomedical Engineering* 33 (4): 545–53.
St. Laurent, Robert, and Jacques Larochelle (1994). 'The Cooling Power of the Pigeon Head.' *Journal of Experimental Biology* 194: 329–39.

MOVING VIOLATIONS

Historically, Europe's washing machines tended to walk across a room, while America's did not. Daniel Conrad and Werner Soedel explained why, in a study called 'On the Problem of Oscillatory Walk of Automatic Washing Machines'. Conrad and Soedel, based at the School of Mechanical Engineering at Purdue University in West Lafayette,

Indiana, published their work in 1995 in the *Journal of Sound and Vibration*. Their explanation has been recognized by authority figures for its power to inspire youths.

The fear of ambling machinery resonated with modern times. One could feel it in the 1995 Japanese science-fiction film *Mechanical Violator Hakaider*. Critic Jason Buchanan later described what happens once the title character, a cyborg, is loosed upon the land: 'Once Hakaider sets on the path of destruction there is little that can be done to stop him from destroying all of Jesus Town.'

Washing machines of that era sometimes contained frightful things. A 1993 detective thriller called *Vortice Mortale* (English title: *The Washing Machine*) cinematically depicted a dismembered man inside an Italian unit.

Conrad and Soedel eschewed the sensational, restricting themselves to the engineering basics. 'The problem of walk in automatic washing machines is becoming more and more of interest to appliance manufacturers', they wrote. 'The current trend is towards lightweight plastic and composite components. The reduction of mass associated with these changes in materials increases the possibility that a washing machine will walk.'

In washing machines, the propensity to waddle is the consequence of a particular design choice. While steadfast American machines rotated their dirty clothes about a vertical axis, European designs typically made the internal machinery twirl around horizontally.

Conrad and Soedel saw this as a mechanical and business blunder. They wrote: 'The horizontal axis washer has innate unbalance problems associated with the design. This unbalance can typically create a force in excess of 19 kilonewtons during the spin cycle.'

Four years after the publication of 'On the Problem of Oscillatory Walk of Automatic Washing Machines', two officers at the US Military Academy in West Point, New York, used it as a major source for their paper 'Basic Vibration Design to Which Young Engineers Can Relate: The Washing Machine'.

Lt. Col. Wayne Whiteman and Col. Kip Nygren pointed out that 'virtually every campus has laundry facilities for students. Most students are therefore familiar with the unwanted vibrations that occur when an unbalance of clothes accumulates during the spin cycle.'

Young engineers thrill at bad vibrations. Keying on that, White-man and Nygren sketched out, in terms designed to resonate with their audience, the story of how to prevent oscillatory walk. These terms are lyrical, if you are a certain type of engineer, and perhaps someone will use them in a hip-hop hit: Mass of the Entire Machine; Mass of Inner Housing and Rotating Drum; Mass of Unbalanced Clothes; Coefficient of Static Friction with Floor; Radial Distance to Unbalanced Clothes; Spin Speed; Suspension Spring Constant; Suspension Damping Ratio.

Conrad, Daniel C., and Werner O. Soedel (1995). 'On the Problem of Oscillatory Walk of Automatic Washing Machines.' *Journal of Sound and Vibration* 188 (3): 301–14.

Whiteman, Wayne E. and Kip P. Nygren (1999). 'Basic Vibration Design to Which Young Engineers Can Relate: The Washing Machine.' Paper presented at the annual meeting of the American Society for Engineering Education, Charlotte, N.C., 20–23 June, session 3268.

THE THREAT OF THE ROBO-TOASTER

Which kind of robot will be the first to arise and smite us? A study called 'Experimental Security Analysis of a Modern Automobile' suggests we keep an eye on the family car.

Written by Karl Koscher and a team of ten other researchers at the University of Washington and at the University of California, San Diego, the paper was presented at the 2010 IEEE (Institute of Electrical and Electronics Engineering) Symposium on Security and Privacy, in Berkeley, California.

Unlike the mindless jalopies of the past, the paper points out, 'Today's automobile is no mere mechanical device, but contains a myriad of computers.'

This myriad has powers to do good things for us humans, as well as bad things to us. Already, in some cases, the microchip hordes quietly, beneficently take control from the driver. The Lexus LS460 luxury sedan can automatically parallel-park itself. Many General Motors cars will soon have what the study calls 'integration with Twitter'. Other abilities are just around the corner.

The team's goal was to look past the goodness, and see how hard it would be to cause trouble.

Limiting themselves to the here and now ('we concern ourselves solely with the vulnerabilities in today's commercially available automobiles'), they tell, in professionally dull, let's-remember-we're-engineers fashion, how they conducted an experimental reign of terror: 'We have demonstrated the ability to systematically control a wide array of components including engine, brakes, heating and cooling, lights, instrument panel, radio, locks, and so on. Combining these we have been able to mount attacks that represent potentially significant threats to personal safety. For example, we are able to forcibly and completely disengage the brakes while driving, making it difficult for the driver to stop. Conversely, we are able to forcibly activate the brakes, lurching the driver forward and causing the car to stop suddenly.'

They played other sorts of dangerous tricks, too, with the greatest of ease. At their behest, speeding cars shot windscreen-washing fluid continuously, popped the trunk, blared the horn, and, in a grim sense, had a high old time.

The study focuses on cars. But indirectly, it foresees the day when our very toasters and teapots might turn or be turned against us. On that question there is mystery, if not much dread, in part because there's little publicly available research about the threat of hijackable household appliances. In 1996, security experts based partly at the RAND Corporation wrote a report called 'Information Terrorism: Can You Trust Your Toaster?' Mainly they (1) recommend hiring lots of 'information warriors', but warn that (2) law enforcement agencies sometimes squabble, and so (3) 'information terrorists' could inflict damage 'in the time it takes to argue about whose job it is to respond'. More mundanely, Austin Houldsworth of the Royal College of Art in London created what may be the world's most dangerous teapot, and the quickest. Houldsworth tells how it works: 'The heating elements within the kettle contain thermite, which ... burns at 2500 degrees.' (See it in action at http://vimeo.com/5043742.)

Koscher, Karl, Alexei Czeskis, Franziska Roesner, Shwetak Patel, Tadayoshi Kohno, Stephen Checkoway, Damon McCoy, Brian Kantor, Danny Anderson, Hovav Shacham, and Stefan Savage (2010). 'Experimental Security Analysis of a Modern Automobile.' Paper presented at the 2010 IEEE Symposium on Security and Privacy, Berkeley, Calif., 16–19 May, http://www.autosec.org/pubs/cars-oakland2010.pdf.

DRESS STRESS ENGINEERING

Charles Seim is project engineer of the Gibraltar Bridge, the somewhat whimsically proposed megagigantic structure that would join Spain and Morocco, spanning a distance of five miles across the Strait of Gibraltar. He prepared for this perilous task, early in his career, by writing a report entitled 'Stress Analysis of a Strapless Evening Gown'.

'Effective as the strapless evening gown is in attracting attention', Seim wrote in 1956, 'it presents tremendous engineering problems to the structural engineer. He is faced with the problem of designing a dress which appears as if it will fall at any moment and yet actually stays up with some small factor of safety.'

The study includes two technical drawings. The first is a front view of the torso of a woman wearing a strapless gown. It will be familiar in kind, if not in all its details, to anyone who has studied physics on any level.

Seim's prose fleshes out the fine points. Here is a typical passage: 'If a small elemental strip of cloth from a strapless evening gown is isolated as a free body in the area of plane A in Figure 1, it can be seen that the tangential force F1 is balanced by the equal and opposite tangential force F2. The downward vertical force W (weight of the dress) is balanced by the force V acting vertically upward due to the stress in the cloth above plane A. Since the algebraic summation of vertical and horizontal forces is zero and no moments are acting, the elemental strip is at equilibrium.'

Figure 2 offers a detailed side view of the bust. Seim uses it to illustrate the kind of daunting technical challenge that good engineers relish. His prose brings vivacity to the spare draftsmanship and simple mathematical notation. This is how he introduces the chief difficulty posed by the upper surface of the breast: 'Exposure and correspondingly more attention can be had by moving the dress line from a toward b. Unfortunately, there is a limit stress defined by $S = F/2A$ (A being the area over which the stress acts). Since $F/2$ is constant, if the area A is decreased, the bearing stress must increase. The limit of exposure is reached when the area between b and c is reduced to a value of "danger point".'

Over the past fifty years, Charles Seim's concept of an engineering danger point has inspired many people to see the drama inherent in the analysis of tension, compression, stress, and strain. In 1992, it inspired an homage from jazz harpist and singer Deborah Henson-Conant, a five-movement orchestral composition called *Stress Analysis of a Strapless*

Evening Gown. Henson-Conant performs this technical gem regularly with symphony orchestras. Each time, she wears a well-engineered strapless evening dress, which she loves. Her hope is to keep it up.

Seim, Charles E. (1956). 'Stress Analysis of a Strapless Evening Gown.' *The Indicator* November.

PECKER BANG ANALYSIS

While others tried to build a better computer or teapot or mousetrap, Julian F. V. Vincent, Mehmet Necip Sahinkaya, and W. O'Shea of the department of mechanical engineering at the University of Bath tried to build a better hammer. Unlike most previous hammer smiths, they studied woodpeckers. Why? Because to mechanical engineers, when they are in a certain frame of mind, a woodpecker is nature's finest version of a hammer.

The trio published 'A Woodpecker Hammer' in a scholarly journal with the unwieldy name *Proceedings of the Institution of Mechanical Engineers, Part C, Journal of Mechanical Engineering Science*.

There they begin with a nod to the Ig Nobel Prize-winning research of Dr Ivan Schwab of the University of California-Davis School of Medicine, who in 2002 wrote a monograph that explains why woodpeckers don't get headaches. Schwab was fascinated by the mechanical properties of the woodpecker's head – especially why its brain doesn't homogenize during all that pummelling, and why its eyes don't pop out of their sockets. The Bath scientists take a more holistic approach. They explore how the bird's entire body, from head to toes, feathers included, effectively function as a simple mechanical tool for pounding wood.

Vincent, Sahinkaya, and O'Shea examined a green woodpecker (*Picus viridis*) that was in the terminal state known as 'road kill'. They measured the remains using old-fashioned methods and also with X-ray equipment, thus determining the values for several parameters: head mass, body mass, and the relative lengths of the parts. Using these, and also video of a living, pecking woodpecker of similar size, the scientists estimated the bird's head inertia, body inertia, neck stiffness, neck damping, and body spring stiffness. They wrote equations to describe a woodpecker's motions as it moves through all phases of the drum-drum-drum-on-wood cycle. To keep the mathematics fairly simple, there were a few engineering simplifications. The woodpecker's vertebrae and neck tendons together behave as a spring. The tree is, essentially, a stiff spring with a damper.

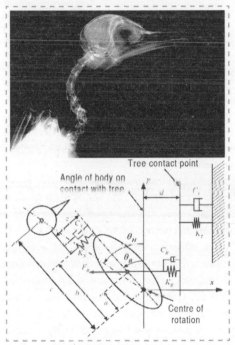

X-ray of a road-kill green woodpecker (top); schematic of woodpecker at work (bottom)

The study proudly proclaims an intended payoff from this research: 'One of the reasons for studying the woodpecker was to derive a design for a lightweight hammer. It was reasoned that the woodpecker is a bird, therefore has to fly and therefore is constructed as light as possible. The mechanism, which has emerged as a result of the model reported here – momentum transfer from body to head of the woodpecker – has been used in the design of a novel hammer [in which a] rotating crank is connected by means of a rod to the casing, so that the motor plus its mounting oscillates about a central pin.'

Vincent, Sahinkaya, and O'Shea say their original intent was to use this hammer in space exploration, 'where it has no net inertia until it comes in contact with an object'. But its first use, they confide, probably will be in dentistry.

Vincent, Julian F. V., Mehmet Necip Sahinkaya, and W. O'Shea (2007). 'A Woodpecker Hammer.' *Proceedings of the Institution of Mechanical Engineers, Part C, Journal of Mechanical Engineering Science* 221 (10): 1141–7.

THE PHYSICS OF SKULKING AND FALLING CATS

Cats may skulk, and cats may fall – but no matter what they do, cats must obey the laws of physics. Scientists have tried repeatedly to figure out how they manage to do it.

At the extreme, physicists analysed what happens to a dropped cat. That's a cat in free-fall, a cat hurtling earthwards with nothing but kitty cunning to keep it from crashing.

In 1969, T. R. Kane and M. P. Scher of Stanford University published their monograph 'A Dynamical Explanation of the Falling Cat Phenomenon'. It remains one of the few studies about cats ever published in the *International Journal of Solids and Structures*. Kane and Scher explain: 'It is well known that falling cats usually land on their feet and, moreover, that they can manage to do so even if released from complete rest while upside-down ... numerous attempts have been made to discover a relatively simple mechanical system whose motion, when proceeding in accordance with the laws of dynamics, possesses the salient features of the motion of the falling cat. The present paper constitutes such an attempt.'

And what an attempt it is!

Kane and Scher neither lifted nor dropped a single cat. Instead, they created a mathematical abstraction of a cat: two imaginary cylinder-like chunks, joined at a single point so the parts could (as with a feline spine) bend, but not twist. When they used a computer to plot the theoretical bendings of this theoretical falling chunky-cat, the motions resembled what they saw in old photographs of an actual falling cat. They conclude that their theory 'explains the phenomenon under consideration'.

In 1993, a professor at the University of California, Santa Cruz, applied some heavier-duty mathematics and physics tools to the same question. Richard Montgomery's study, called 'Gauge Theory of the Falling Cat', leaps and bends across twenty-six pages of a mathematics journal. Then it mutters that 'the original solutions of Kane and Scher [are] both the optimal and the simplest solutions'.

But cats rarely fall from the sky. More commonly, they skulk. On the ground. And skulking cats are just as provocative, to a physics-minded scientist, as plummeting cats.

In 2008, Kristin Bishop of the University of California, Davis, together with Anita Pai and Daniel Schmitt of Duke University in North Carolina, published a report called 'Whole Body Mechanics of Stealthy Walking in Cats', in the journal *PLoS One*.

They studied six cats, three of which 'were partially shaved and marked with contrasting, non-toxic paint to aid in kinematic analysis'. They discovered 'a previously unrecognised mechanical relationship' between 'crouched postures', 'changes in footfall pattern', and the amount of energy needed to produce those crouched-posture footfall patterns.

Cats that intend to skulk, in Bishop, Pai, and Schmitt's view, are hemmed in by the laws of the physical universe. They must make 'a tradeoff between stealthy walking', which uses a lot of energy, and plain old, energy-efficient cat-walking.

Kane, T. R., and M. P. Scher (1969). 'A Dynamical Explanation of the Falling Cat Phenomenon.' *International Journal of Solids and Structures* 5: 663–70.

Montgomery, Richard (1993). 'Gauge Theory of the Falling Cat.' *Fields Institute Communications* 1: 193–218.

Bishop, Kristin L., Anita K. Pai, and Daniel Schmitt (2008). 'Whole Body Mechanics of Stealthy Walking in Cats.' *PLoS One* 3 (11): e3808.

THREE

DOGS, COWS, CATS, AND SO FORTH

IN BRIEF

'THE COW WITH ZITS'
by Walter J. Pories (published in *Current Surgery*, 2001)

Some of what's in this chapter: Frightening a cow in the forties • Motivating the domestic fowl • Yawning contagiously, or not, with fellow tortoises • Lying down and standing up, lying down and standing up, and mooing • Following Fish on fish, and Fish on trees • Roaring with or without a Klipsch Heresy Speaker placed two hundred metres away in the jungle • Bollocks for dogs; rolls for kitties • Collecting lizards from the sky • Towing and showing naked Russian swimmers in place of dolphins • Macaque upchuck

CATTLE RUSTLING

What can be learned with a cat, a cow, and a paper bag? This is not a moot question. To raise dairy cows can be intellectually challenging, in addition to being hard physical work. Every dairy farmer knows this, although it may be news to a small number of milk-guzzling, cheese-chomping city dwellers.

Fordyce Ely and W. E. Petersen wanted to understand why some cows spew their milk. This was in the early 1940s. Much of the world was at war, which may explain why Ely and Petersen's report, titled 'Factors Involved in the Ejection of Milk', made only a little splash when it was published in 1941 in the *Journal of Dairy Science*. Ely was based at the Kentucky Agricultural Experiment Station, and Petersen at a similar institute in Minnesota. Together they made history, using the aforementioned items – a cat, a cow, and a paper bag.

Ely and Petersen set out to address a nagging dilemma. 'Cows which habitually "let down" or "hold up" their milk are common in all herds. Several theories have been advanced to explain the physiological processes involved, but each has been found at fault in some regard.'

In search of the truth, they conducted an experiment. The details deal with complex aspects of the nervous system as it relates to the physiology of bovine udders, but I will concentrate here on just one aspect. Here is the pertinent passage from Ely and Petersen's report: 'It was thought that there might be a difference in the response of the two halves of the udder as measured by the rate of ejection of milk if the cow was severely frightened. Accordingly, [the cow] was systematically frightened as the mechanical milker was attached. Frightening at first consisted in placing a cat on the cow's back and exploding paper bags every ten seconds for two minutes. Later the cat was dispensed with as unnecessary.' So far as I have been able to determine, this experiment was conducted only that one time.

Other scientists tried to startle human beings. Often, they succeeded.

D. N. May of the University of Southampton, UK, carried out one such experiment. In a 1971 report, he writes, '[My] result contradicts a previous finding with animals and suggests that sonic booms are likely to be more startling in quiet environments than noisy ones.'

Not long afterwards, J. S. Lukas at the Stanford Research Institute exposed some sleeping Californians to recorded aircraft sounds and simulated sonic booms. He found that anyone over the age of eight is likely to notice.

Meanwhile, researchers at the Karolinska Institute in Stockholm used real jet planes to produce real sonic booms. They discovered that when you do this at 4 a.m., it awakens the majority of Swedish adults.

Ely, Fordyce, and W. E. Petersen (1941). 'Factors Involved in the Ejection of Milk.' *Journal of Dairy Science* 3: 211–23.

May, D. N. (1971). 'Startle in the Presence of Background Noise.' *Journal of Sound and Vibration* 17 (1): 77–78.

Lukas, Jerome. S. (1972). 'Awakening Effects of Simulated Sonic Booms and Aircraft Noise on Men and Women.' *Journal of Sound and Vibration* 20 (4): 457–66.

Rylander, R., S. Sörensen, and K. Berglund (1972). 'Sonic Boom Effects on Sleep: A Field Experiment on Military and Civilian Populations.' *Journal of Sound and Vibration* 24 (1): 41–50.

MAGNETIC CHICKENS

Progress comes slowly on the question 'Why does the chicken cross the road?' But come it does. The answers (for there seem to be many) strut in jerkily, from different directions. A new study explains that magnetic fields play some sort of role, at least sometimes, in chickens' decisions to navigate hither or yon.

The study has a title that seems swiped from a children's book: 'The Magnetic Compass of Domestic Chickens, *Gallus gallus*'. Published in the *Journal of Experimental Biology*, a venue that generally does not cater to youngsters, it adds flesh and feathers to the sketchy picture revealed in an earlier report entitled 'Chickens Orient Using a Magnetic Compass'.

Many cultures wonder about the chicken navigation mystery. Fittingly, the 'magnetic compass' research team is international. Its members – Wolfgang Wiltschko, Rafael Freire, Ursula Munro, Thorsten Ritz, Lesley Rogers, Peter Thalau, and Roswitha Wiltschko – work, variously, in Germany, at J. W. Goethe-Universität Frankfurt; in Australia at the University of New England in Armidale and at the University of Technology, Sydney; and in the US, at the University of California, Irvine.

They conducted the experiments all in a single land: Australia. The chickens, which were domestic to that nation, had to track down a red ball that had been shown to them but was then moved. The scientists generated a magnetic field that, they hoped, would monkey with the chickens' orientation. The chickens, in their quest for the red ball, acted as if they were under the influence of a monkeyed-with magnetic field.

Thus the scientists' conclusion: magnetic fields matter to chickens. This presumably is true of chickens elsewhere, although the report is not explicit on the point.

What do magnetic fields do for the chicken? They 'facilitate orientation within the home range', say the researchers. In more specific terms: 'Tests in magnetic fields with different intensities revealed a functional window around the intensity of the local geomagnetic field, with this window extending further towards lower than higher intensities.'

What the report hints, but does not quite say, is that chickens don't seem to rely heavily on magnetism. Are the chickens capable of more, for instance, of making more intelligent use of what they perceive? Has our (and to some extent their) civilization withered the birds' reliance on the Earth's magnetosphere? The report is mum on these questions.

The traditional question of why a chicken crosses a road stands, at best, partially answered. Perhaps the most likely to solve it is Professor Ian J. H. Duncan, formerly of the Poultry Research Centre in Edinburgh, and now chair in animal welfare at the department of animal and poultry science at the University of Guelph, in Canada.

In 1986, Duncan and a colleague presented a paper at the winter meeting of the Society for Veterinary Ethology, in London. The title: 'Some Investigations into Motivation in the Domestic Fowl'.

> Working for a dustbath: are hens increasing
> pleasure rather than reducing suffering?
>
> Tina M. Widowski *, Ian J.H. Duncan
>
> Department of Animal and Poultry Science, University of Guelph, Guelph, Ontario, Canada N1G 2W1

Duncan appears to be methodical about his scratchings into fowl motivation. In 2000 he co-authored 'Working for a Dustbath: Are Hens Increasing Pleasure Rather Than Reducing Suffering?' Pullet-road-crossing enthusiasts can hope that one day Duncan will confront, directly, the question of questions.

Wiltschko, Wolfgang, Rafael Freire, Ursula Munro, Thorsten Ritz, Lesley Rogers, Peter Thalau, and Roswitha Wiltschko (2007). 'The Magnetic Compass of Domestic Chickens, Gallus gallus.' Journal of Experimental Biology, 210 (13): 2300–10.

Freire, Rafael, Ursula H. Munro, Lesley J. Rogers, Roswitha Wiltschko, and Wolfgang Wiltschko (2005). 'Chickens Orient Using a Magnetic Compass.' Current Biology, 15 (16): R620–21.

Duncan, Ian J. H., and V. G. Kite (1986). 'Some Investigations into Motivation in the Domestic Fowl.' Applied Animal Behaviour Science 18 (3–4): 387–88.

Widowski, Tina M., and Ian J. H. Duncan (2000). 'Working for a Dustbath: Are Hens Increasing Pleasure Rather than Reducing Suffering?' Applied Animal Behaviour Science 68 (1): 39–53.

CONTAGIOUS YAWNING IN THE RED-FOOTED TORTOISE

Scientists know a bit more about contagious yawning – one of science's utter mysteries – thanks to a study called 'No Evidence of Contagious Yawning in the Red-Footed Tortoise Geochelone carbonaria'. The study's authors say their experiments, conducted with seven tortoises, might help eliminate some of the many competing theories as to why humans yawn when they see other humans yawn.

Writing in the journal *Current Zoology*, Anna Wilkinson, Isabella Mand, and Ludwig Huber of the University of Vienna, Austria, and Natalie Sebanz of Radboud University in the Netherlands, share their hopes: 'This study aimed to discriminate between the possible mechanisms controlling contagious yawning by asking whether contagious yawning is present in a species that is unlikely to show empathy or nonconscious mimicry: the red-footed tortoise *Geochelone carbonaria*.'

The researchers say that although tortoises have not been known (by humans) to empathize with or mimic each other, the animals do sometimes respond to things they see around them. That makes the tortoises 'ideal subjects for examining this question'.

The tortoises, whose names are Alexandra, Moses, Aldous, Wilhelmina, Quinn, Esme, and Molly, were old hands at being scientific test subjects. The study notes: 'None of the tortoises were experimentally naïve, but they had never previously been involved in a contagious yawning task or any similar experiment.'

The researchers trained Alexandra to open her mouth whenever they waved a little red square near her head. 'This took 6 months', they write, and 'the resulting behavior appeared highly similar to a naturally occurring tortoise yawn ... The yawn is extremely clear and cannot be mistaken for another behavior.' Alexandra thus became the 'demonstrator', the individual who yawned in plain view of her fellows.

In one experiment, the other tortoises watched as Alexandra yawned a single time. In a second experiment, Alexandra yawned several times in rapid (for a tortoise) succession. In the third and final experiment, the observer tortoises watched videos of a tortoise (a) yawning and (b) not yawning. After reviewing the evidence, the scientists determined that there is 'the suggestion that tortoises do not yawn in a contagious manner'.

The monograph ends with a statement expressing gratitude to their colleagues. Perhaps lacking for warmth, it says, 'Acknowledgements: The authors would like to thank the cold-blooded cognition group at the University of Vienna for their helpful comments.'

The team's slow, careful attention to tortoise yawns led to glory of a sort. Wilkinson, Mand, Huber, and Sebanz were awarded the 2010 Ig Nobel Prize in physiology.

Other scientists have experimented with contagious yawning in other species. I will mention just one: 'Some Comparative Aspects of Yawning in *Betta splendens, Homo sapiens, Panthera leo,* and *Papio sphinx*' by Ronald Baenninger of Temple University in Philadelphia.

'In this report', Baenninger writes, 'I describe observations of yawning by a fish [Siamese fighting fish], by a carnivore [a lion], and by two primate species [mandrills and humans].' The humans watched 'a semiprofessional actor read a passage from *Alice in Wonderland* (the mock turtle's story)'. The other animals watched nonprofessional nonactors of their own species.

As a final note, I first learned of the tortoise-yawning study from Stefano Ghirlanda, himself a 2003 Ig Nobel Prize winner in the category of 'interdisciplinary research' for the study 'Chickens Prefer Beautiful Humans'. The title of his study is, to some extent, self-explanatory and, to some extent, probably not.

Wilkinson, Anna, Natalie Sebanz, Isabella Mand, and Ludwig Huber (2011). 'No Evidence of Contagious Yawning in the Red-Footed Tortoise *Geochelone carbonaria*.' *Current Zoology* 57 (4): 477–84.

Baenninger, Ronald (1987). 'Some Comparative Aspects of Yawning in *Betta splendens, Homo sapiens, Panthera leo,* and *Papio sphinx*.' *Journal of Comparative Psychology* 101 (4) 349 54.

Ghirlanda, Stefano, Liselotte Jansson, and Magnus Enquist (2002). 'Chickens Prefer Beautiful Humans.' *Human Nature* 13 (3): 383–89.

MAY WE RECOMMEND

'DETERMINING THE SMALLEST MIGRATORY BIRD NATIVE TO BRITAIN ABLE TO CARRY A COCONUT'

by Robert Hopton, Steph Jinks, and Tom Glossop (published in the *Journal of Physics Special Topics,* 2010)

This report pertains to King Arthur's postulation in the film *Monty Python and the Holy Grail* that a migratory bird could have transported coconuts from the tropics to Britain. Hopton, Jinks, and Glossop calculate that the only British bird with a chance at succeeding is the white stork. No go, they warn. The stork's cross-sectional area is slightly too low to provide the required amount of lift. The stork would fall short, and King Arthur would be nutless.

THE UPS AND DOWNS OF COWS

A new study called 'Are Cows More Likely to Lie Down the Longer They Stand?' adds to our knowledge of what cows do and why they do it.

Some researchers succumb to temptation – hazarding unprovable guesses as to cows' intentions, motivations, and desires. Five scientists in Scotland, though, took a careful path, methodically measuring a very specific part of the what, and not guessing too wildly at the why.

Bert Tolkamp, Marie Haskell, Fritha Langford, David Roberts, and Colin Morgan, based at the Scottish Agricultural College, published their monograph in the journal *Applied Animal Behaviour Science*. It builds upon a large body of work by other researchers.

Some of the earlier reports have almost-poetical titles. The best in that respect is (in my opinion, at least) a Swedish report called 'Effects of Milking Frequency on Lying Down and Getting Up Behaviour of Dairy Cows'. Its authors, Sara Osterman and Ingrid Redbo of the Kungsängen Research Centre in Uppsala, argue that milking thrice a day – rather than twice – 'contributes to increased comfort in high-producing dairy cows'. The Scottish team focused on questions that stem indirectly from this Swedish study.

Tolkamp, Haskell, Langford, Roberts, and Morgan set out to test two hypotheses – two educated guesses – about the nature of cowhood.

First, they hypothesized that the longer a cow has been lying down, the more likely it soon will stand up. After gathering lots of what-did-the-cows-do data, they report that yes, this is exactly what happens. Generally speaking, you can't keep a good cow down, not for long, not if the cow is healthy.

Their second hypothesis looked at things the other way around. They predicted that the longer a cow has been standing up, the more likely it is to lie down. Here the cows gave them a surprise.

Lying down and getting up movements were analysed in detail and registered in accordance with the following definitions. The lying down movement was divided into two phases.

Lying down, phase 1	Started when the nose was moved in a pendulum movement close to the ground and ended when the cow had one knee on the floor.
Lying down, phase 2	The time it took for the cow to move from one knee on the floor until the lying down movement was completed, i.e. when the cow lies down on one of its two hips.

The getting up movement is just referred to as one movement.

After ruminating over their data, the team decided that no, their expectation was wrong. The truth, they conclude, is that once a cow has stood up, you can't easily predict how soon it will lie down again.

This kind of experiment, if it is to produce trustworthy results, requires a series of careful technical decisions. How many cows should you watch, under what circumstances, and for how long? How can you reliably monitor whether and when each cow has officially stood up or flopped down?

The scientists examined three groups of cows, seventy-three individuals all told. They attached an electronic sensor to each animal, to automatically note and record the cow's ups and downs. They then validated some of the sensors' sensings, by watching video recordings of some of the cows and comparing what they saw with what the sensors had said.

Some uncertainties persist. 'The question why some cows had total daily resting times less than half of those achieved by other cows in the same experiment, as well as a large number of other questions', says the report, 'remain to be addressed in future research.'

(Ig Nobel Prize winner Richard Wassersug, he of the tadpole-tasting experiment and the eunuch research, brought this cow work to my attention. Professor Wassersug is a man of wide-ranging curiosity and, apparently, incessant scientific and literary grazing habits.)

Tolkamp, Bert J., Marie J. Haskell, Fritha M. Langford, David J. Roberts, and Colin A. Morgan (2010). 'Are Cows More Likely to Lie Down the Longer They Stand?' *Applied Animal Behaviour Science* 124, (1-2): 1–10.
Osterman, Sara, and Ingrid Redbo (2001). 'Effects of Milking Frequency on Lying Down and Getting Up Behaviour of Dairy Cows.' *Applied Animal Behaviour Science* 70 (3): 167–76.

HOW NOW WARM COW?

You cannot easily ignore the report called 'A Quick and Accurate Estimation of Heat Losses from a Cow', not if you obsess about rapid calculation techniques, or thermodynamics, or at least one cow. The four scientists responsible – Zahid A. Khan, Irfan Anjum Badruddin, G. A. Quadir, and K. N. Seetharamu – are based at universities in India and Malaysia. They infuse their writing, published in the journal *Biosystems Engineering*, with abundant detail and occasionally strained grammar. Their method, they assure us, 'can be used by any user to predict quickly accurate amount of heat loss from a cow'.

Inevitably comes the question: 'Why would someone want to estimate the heat losses from a cow?' Khan, Badruddin, Quadir, and Seetharamu provide an answer in their first paragraph. 'In order to increase milk yield of the cows,' they write, 'it is necessary to cool them.' This amounts to chilling the milk before making it, long before there is any possibility of serving up a cupful.

The title 'A Quick and Accurate Estimation of Heat Losses from a Cow' implies that there is at least one other way to estimate the heat losses from a cow, and that other method suffers from slowness or inaccuracy, or both. The main one until now – the gold standard – was developed by Kifle G. Gebremedhin of Cornell University in New York and Binxin Wu of Tongji University in China.

The Gebremedhin-Wu method certainly is slow. Despite making a simple assumption – that a cow is a cylinder – it requires you to do some tedious calculating. Khan, Badruddin, Quadir, and Seetharamu, in introducing their own method, pooh-pooh the Gebremedhin-Wu way. They say it involves complex computer programming and, moreover, is useless to people 'who do not possess adequate background of heat and mass transfer in addition to the computer programming skill'.

The old method involved measuring or calculating a whole herd of numbers: the cow's weight; the diameter the cow would have if it were a cylinder; the diameter of a typical hair; the fur density and the fur thickness; the ratio of fur to skin surface area; the coefficient of effective radiant area; the coefficient of radiant heat transfer; the radiant emissive coefficient of the skin; the thermal conductivity of the air and, separately, that of the fur layer.

The new way is simpler. You measure or calculate just four things: wetness of the cow; air temperature near the cow; wind speed; and relative humidity.

And then – the great triumph of the method – you look up the answer in a table. Khan et al. have removed most of the tedium by doing the calculations for you. That's why you don't have to do the calculations yourself.

This triumph of simplification is reminiscent of another study, also describing a new method to replace a tedious old one, and also performed in India. In 1990, K. P. Sreekumar and G. Nirmalan of

Kerala Agricultural University published a report called 'Estimation of the Total Surface Area in Indian Elephants'. They reaped unexpected dividends twelve years later, when they received an Ig Nobel Prize in the field of mathematics.

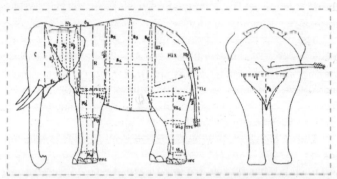

Geometric division of an elephant: view of sites at which body measurements have been taken (left), including 'perineal region base length and altitude' (right)

Khan, Zahid A., Irfan Anjum Badruddin, G. A. Quadir, and K. N. Seetharamu (2006) 'A Quick and Accurate Estimation of Heat Losses from a Cow.' *Biosystems Engineering* 93 (3): 313–23.
Gebremedhin, K. G., and B. X. Wu (2003). 'Characterization of Flow Field in a Ventilated Space and Simulation of Heat Exchange between Cows and Their Environment.' *Journal of Thermal Biology* 28 (4): 301–19.
Sreekumar, K. P., and G. Nirmalan (1990). 'Estimation of the Total Surface Area in Indian Elephants (*Elephas maximus indicus*).' *Veterinary Research Communications* 14 (1): 5–17.

FISHES (MANY OF THEM IN SCHOOLS) ON FISH

What do Fishes know? Quite a lot, it turns out. Here are some studies done by investigators who are, or at least are named, Fish.

> *Fish on Flatfish.* E. Brainerd, B. Page, and F. Fish, 'Opercular Jetting During Fast-Starts by Flatfishes' (published in the *Journal of Experimental Biology*, 1997). Fish and his friends report that: 'When attacked by predators, flatfishes perform fast-starts that result in a rapid take-off from the ocean bottom on which they lie ... [W]e simulated fast-starts using a physical model in which a dead flounder was pulled upwards with an acceleration of 95 meters per second....'

Fish on Whitefish. Sylvan M. Fish, et al. 'Epidemic of Febrile
    Gastroenteritis Due to *Salmonella java* Traced to Smoked
    Whitefish' (published in the *American Journal of Public
    Health and the Nation's Health*, 1968).

Fish on Fish Oil. S. C. Whitman, J. R. Fish, et al. 'N-3 Fatty Acid
    Incorporation into LDL Particles Renders Them More
    Susceptible to Oxidation in Vitro But Not Necessarily
    More Atherogenic in Vivo' (published in *Arteriosclerosis
    and Thrombosis*, 1994).

Fish on Whales. Frank E. Fish and Juliann M. Battle. 'Hydro-
    dynamic Design of the Humpback Whale Flipper' (pub-
    lished in the *Journal of Morphology*, 1995).

Fish on Seals. Frank E. Fish, S. Innes, and K. Ronald. 'Kinemat-
    ics and Estimated Thrust Production of Swimming Harp
    and Ringed Seals' (published in the *Journal of Experimen-
    tal Biology*, 1988).

Fish on Platypus. F. E. Fish, R. V. Baudinette, et al. 'Energetics
    of Swimming by the Platypus *Ornithorhynchus anatinus*:
    Metabolic Effort Associated with Rowing' (published in
    the *Journal of Experimental Biology*, 1997).

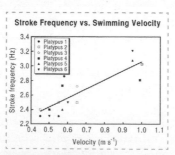

**Fish's stroke frequency vs. swimming
velocity in six platypuses**

Fish on Ducks. Terrye L. Aigeldinger and Frank E. Fish, et al.
    'Hydroplaning by Ducklings: Overcoming Limitations
    to Swimming at the Water Surface' (published in the
    Journal of Experimental Biology, 1995).

Fish on Muskrats. F. E. Fish. 'Mechanics, Power Output and

Efficiency of the Swimming Muskrat (*Ondatra zibethicus*)'
(published in the *Journal of Experimental Biology* 1984).

Shark on Yeast. S. A. Johnston, P. Q. Anziano, K. Shark, et al.
'Mitochondrial Transformation in Yeast by Bombardment
with Microprojectiles' (published in *Science*, 1988).

Fish on Fungi. L. I., Lulinich, E. D. Bershadskaia, N. G. Fish,
et al., 'Use of the Coagglutination Reaction of Yeast-Like
Candida maltosa Fungi for Detecting Fimbriae in Intestinal Bacteria' (published in Russian in *Zhurnal Mikrobi-
ologii, Epidemiologii, i Immunobiologii*, 1988).

Fish on Trees. J. M. Friedman and R. D. Fish. 'The Use of Probability Trees in Genetic Counselling' (published in *Clinical
Genetics*, 1980).

The Fishes, as a group, do not enjoy any special standing within
the scientific and educational communities. Perhaps any little attention
and appreciation we give them will help change that.

THE PRIDE OF THE PRIDE

Lion-roaring competitions used to be private, simple affairs, organized entirely by lions, without spectators. That changed in the early
1990s, when Karen McComb, Jon Grinnell, Craig Packer, and Anne
Pusey realized they could use technology – loudspeakers, amplifiers,
and sometimes a stuffed artificial lion – to stage-manage some lion-
roaring contests, and to document those ginned-up events on video.

The foursome wanted to know: when lions hear other lions roar,
what do they do?

McComb was based at the University of Cambridge, Grinnell at the
College of Wooster and at the University of Minnesota, and Packer and
Pusey at the University of Minnesota. The roaring contests, however,
were held in Tanzania.

The researchers set up loudspeakers in the jungle, booming out
recordings they had made of one, two, or three lions roaring simultaneously.

In a series of reports in the journal *Animal Behaviour*, they detail
what happened. First, they give some context. Lion society is organized in prides – groups of a few females, even fewer males, and some

offspring. There are also quite a few nomadic males, who (as the old joke goes) have no pride.

The monograph 'Roaring and Social Communication in African Lions' is all about masculine roaring. Groups of males in their own territory listening to recorded, amplified roars, generally roared back, and often walked towards the loudspeaker. Nomadic males heard the same recordings, but, being uninvited guests, they always stayed silent and kept to themselves.

A monograph called 'Roaring and Numerical Assessment in Contests between Groups of Female Lions' tells how 'recordings of single females roaring and groups of three females roaring in chorus were played back to simulate the presence of unfamiliar intruders'.

Female lions, we're told, 'deliver their roars in bouts which generally last less than a minute and consist of several soft introductory moans, a series of full throated roars and a terminating sequence of grunts. When pride members roar together the bout is delivered in chorus, one individual initiating and others joining in as the bout progresses by adding their roars in an overlapping fashion.'

The females who listened to recordings sometimes responded, but sometimes didn't. It seemed to depend, more or less, on how many companions were with them, and on how many voices were evident in the recording. Some walked towards the loudspeaker. Some 'attempted to recruit absent pride-mates to the contest by roaring'. The study says: 'On nearly half of these occasions companions joined them at the playback site within an hour.'

You will perhaps want some backstage flavour of the staged events, which officially are called 'controlled artificial contests'. *Voilà*: 'A single bout of roaring lasting 25-55 seconds was played 30 minutes prior to dusk using a Panasonic SV-250 Digital Audio Tape Recorder, an ADS Pl20 amplifier and a Klipsch Heresy Speaker placed at 200 meters from the subjects (as measured on a Land Rover odometer) ... Available vegetation was used to conceal the loudspeaker'.

Who Roars?

The 12 coalitions of resident males gave an average ± SE of 0.99 ± 0.19 roar bouts per lion per h over the period of observation. In contrast, none of the six coalitions of nonresident males observed ever roared (Table 1; Mann–Whitney U test: $U=66$, $N_1=12$, $N_2=6$, $P<0.005$). Roaring was thus confined to resident males.

Are Roarers Willing to Escalate?

All 11 resident coalitions approached the loudspeaker when they were challenged with unfamiliar males and

Discussion of 'Results' in Grinnell and Mc-Comb (2000)

Grinnell, Jon, and Karen McComb (2001). 'Roaring and Social Communication in African Lions:
The Limitations Imposed by Listeners.' *Animal Behaviour* 62 (1): 93–98.
McComb, Karen, Craig Packer, and Anne Pusey (1994). 'Roaring and Numerical Assessment
in Contests Between Groups of Female Lions, *Panthera leo.' Animal Behaviour* 47 (2): 379–87.
Grinnell, Jon, Craig Packer, and Anne E. Pusey (1995). 'Cooperation in Male Lions: Kinship,
Reciprocity or Mutualism?' *Animal Behaviour* 49 (1): 95–105.

AN IMPROBABLE INNOVATION

'SURGICAL METHOD AND APPARATUS FOR IMPLANTATION OF A TESTICULAR DEVICE'

a/k/a Neuticles, artificial replacement testicles for dogs available in three sizes and three degrees of firmness, by Gregg A. Miller (US Patent no. 5,868,140, granted 1999 and honoured with the 2005 Ig Nobel Prize in medicine)

CAT ROLLER

Domestic cats roll. Oh, they roll and roll and roll – not constantly, but often enough that the behaviour eventually caught the attention of scientists. In 1994, Hilary N. Feldman, of Cambridge University's sub-department of animal behaviour, did a formal study of the phenomenon. Feldman's monograph, entitled 'Domestic Cats and Passive Submission', appeared in the journal *Animal Behaviour*.

Other scientists had made little leaping swats at the question. Feldman commends J. M. Baerends-Van Roon and G. P. Baerends' book *The Morphogenesis of the Behaviour of the Domestic Cat*, and also L. K. Corbett's University of Aberdeen PhD thesis, 'Feeding Ecology and Social Organization of Wildcats (*Felis silvestris*) and Domestic Cats (*Felis catus*) in Scotland'. Both came out in 1979, marking that year as the previous high point in cat-rolling scholarship.

But Baerends-Van Roon, Baerends, and Corbett only glanced at rolling. Feldman focused on it, and spent six months observing 'two groups of semiferal cats kept in a large outdoor enclosure'.

Rolling, by Feldman's definition, 'involved an individual cat rolling onto its back, with forepaws held cocked, often with the legs splayed and abdomen exposed ... The exposed position was sometimes held for several minutes and was assumed repeatedly in several instances. This was performed in front of another cat in the majority of cases

(79%), and often the rolling animal would approach rapidly and perform the action before any response to the approach was observed.'

The big question of interest, going into this, was the extent to which 'each cat was equally likely to roll to any other individual' versus the extent to which each cat was not. These were adults. The report specifies that 'kitten behaviour was not examined'.

Over the course of the half year, Feldman observed 175 rolls, of which 138 'had an obvious recipient'.

Females rolled mostly while they were in heat. Adult females rolled almost exclusively for adult males. Younger females went mostly for old guys, too, but occasionally rolled for young males or for females.

Males rolled 'throughout the year'. Feldman writes that 'a substantial proportion [61%] of the rolling behaviour was performed by males, and most of this male-initiated activity was directed towards other males'.

Young males rolled towards adults, but the reverse almost never happened. The adults would 'ignore or tolerate the younger males' presence', suggesting to Feldman 'that rolling may act as passive submission and inhibits the development of overt aggression'.

'Both adult and juvenile males rolled ... towards adult females. As with female rolls, it is likely that these rolls were performed in the context of mating, as they occurred when females were displaying other oestrus-related behaviour (e.g. lordosis [exaggerated curving of the spine], erratic running, treading).'

In summary: 'Rolling behaviour in domestic cats appears to have two functions. Females roll primarily in the presence of adult males ... demonstrating a readiness to mate.' But 'males roll near adult males as a form of subordinate behaviour'.

This 'phenomenon of passive submission', Feldman muses, 'may have relevance for a similar behaviour between pet cats and their owners'.

Feldman, Hilary N. (1994). 'Domestic Cats and Passive Submission.' *Animal Behaviour* 47 (2): 457–59.
Baerends-Van Roon, J. M., and G. P. Baerends (1979). *The Morphogenesis of the Behaviour of the Domestic Cat.* Amsterdam: North-Holland Publishing.
Corbett, L. K. (1979). 'Feeding Ecology and Social Organization of Wildcats (*Felis silvestris*) and Domestic Cats (*Felis catus*) in Scotland.' PhD thesis, University of Aberdeen.

THE LIZARDS THAT FELL TO EARTH

The Bible tells of frogs that fall from the sky. Biologists, on the other hand, tell of lizards that fall from trees.

The biologists – William Schlesinger, Johannes Knops, and Thomas Nash – recount in great detail how they discovered an unsuspected truth about lizards. Their study 'Lizardfall in a California Oak Woodland', published in the journal *Ecology*, is a blow to the reputation of a species once admired for its surefootedness. It's the story of the reptiles' ungraceful fall into the abyss – in this case a plastic bucket – and of the detectives who documented that fall.

Western fence lizards spend a lot of time in trees, walking up and down the branches. But when running after insects or away from predators, say Schlesinger, Knops, and Nash, they frequently lose their grip.

Though based at Duke University in North Carolina and at Arizona State University, the team went a-bucketing far from home, on a southeast-facing oak woodland slope in Monterey County, California. There they set big plastic tubs under the trees to see what might come their way.

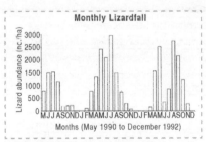

N.B. Lizardfall plummets in December

Lizards are what came their way, for the most part. Bucketloads of lizards. Bucketloads of leaping lizards one might say – but that would be enhancing the facts.

This was no quick overnight stakeout. For nearly three years, beginning early in 1990, Schlesinger et al. set their buckets beneath forty trees, returning monthly to peek inside. In a rousing, earthy passage, they describe a moment of realization, and the effect it had on the investigation: 'When we realized that lizards could not escape over the 43-centimeter sidewalls of the collectors, we began to keep records of lizardfall in May 1990. (Lizards could not climb into the buckets, which slope outwards slightly from bottom to top and which were anchored in the field with short metal stakes after December 1991.) In the summer of 1991 we increased the frequency of our collections to every 2 weeks to prevent the death of captured lizards by dehydration, and in April 1991 we began a protocol of toe-clipping so we could record the recapture of fallen lizards.'

All told, they collected hundreds of fallen western fence lizards. This shattered the animals' reputation for surefootedness, which had been enshrined, for nearly two years, in B. Sinervo and J. B. Losos's 1991 deadpan report 'Walking the Tight Rope: Arboreal Sprint Performance among Sceloporus occidentalis Lizard Populations'.

As documented by Schlesinger et al., the life of a fallen lizard can be grim. The recidivist rate is high. In police-blotter style, the report says: 'Thirty-three percent of the recaptured lizards were found under the same tree as their previous fall.'

And one case was heartbreaking. It's mentioned in a single, plain sentence: 'One particularly clumsy individual was captured five times (in four different collectors) between 20 May 1991 and 16 July 1991, when it was found dead in a collector.'

Schlesinger, William H., Johannes M. H. Knops, and Thomas H. Nash (1993). 'Arboreal Sprint Failure: Lizardfall in a California Oak Woodland.' *Ecology* 74: 2465–67.

INFESTIGATED

Hopes are safer than aspirations, as regards small insects keeping their proper place. Hopes do not by themselves cause an infestation, in the head of a human being, of gnats, midges, anthomid flies, *Collembola*, and wasps parasitic upon the flies. Aspirations can, and sometimes do. This fact slowly, then suddenly, dawns on anyone who reads a report called '"Myiasis" Resulting from the Use of the Aspirator Method in the Collection of Insects', which was published in the journal *Science* in June 1954.

The author, Paul D. Hurd Jr, of the University of California, Berkeley, begins with two paragraphs of impersonal description, written in a rather passive voice. There we learn that 'the aspirator, an apparatus generally designed to collect insects by suction, consists of a vial into which is fitted, by means of a stopper, two pieces of copper tubing, one of which is directed toward the insect and the other is attached to a length of rubber tubing, which during use is placed in the operator's mouth. Across the end of the copper tubing leading to the operator's mouth a fine mesh brass screen is secured. This, of course, is to prevent the aspirated insects being drawn out of the vial and yet provide a free airway between the insect being aspirated and the operator.'

In the third paragraph, in an instant, things perk up. It says: 'Approximately two months after the completion of the past summer's work at Point Barrow [Alaska], I became ill. During the week following the onset of illness four major groups of insects (*Coleoptera, Collembola, Diptera, Hymernoptera*) were passed alive from the left antrum of the sinus.'

The rest of the report maintains this lively tone. It also supplies factual detail, in abundance. Which specific representatives of those four groups of insects emerged from the author's sinus? He supplies this data: 'three adult rove beetles (Staphylinidase), *Micralymna brevilingue* Schiødte; 13 fungus gnat larvae (Mycetophilidae), *Boletina birulai* (Lundstrom); three egg parasite wasps (Mymaridae), *Mymar* sp.; and about 50 springtails (Collembola), *Isotoma olivacea* Tullberg.'

'Myiasis' means infestation. This particular infestation had plenty of time to get a start, and then to fulfil its potential. Hurd aspirated insects for four to six hours every day over the course of an entire summer. Summers in Alaska are shorter than summers in the middle latitudes, but they are long enough for nature to take its full course.

Hurd was given pause. 'I would like to suggest, he writes, 'that those persons who utilize this apparatus so modify it that the flow of air will not be toward the operator's mouth.'

In the report's final paragraph, Hurd drops his restraint, hinting that the tale – and his emotions – are deeper than he's let on.

'It is almost unbelievable', he writes, 'that the insects should have undergone several stages of their metamorphosis within the sinuses.'

Hurd, Paul D., Jr (1954). '"Myiasis" Resulting from the Use of the Aspirator Method in the Collection of Insects.' *Science* 119 (3101): 814–15.

EXPOSURE AT SEA

'Courtship Behaviour of Ostriches Towards Humans Under Farming Conditions in Britain' is the title of a scientific study written by Charles Paxton and three colleagues. In 2002, when I informed Paxton that his team would be awarded that year's Ig Nobel Prize in biology, he took the news matter-of-factly. 'I'm not surprised to be getting this telephone call,' he said, 'but I expected that if I ever won an Ig Nobel Prize, it would be for my work with sea monsters.'

Paxton and two other colleagues, Erik Knatterud and Sharon Hedley, published a study about sea monsters in 2005 that promised to change the way scientists look at the subject. Paxton and Hedley are at St Andrews University in Scotland; Knatterud is based in Stavsjoe, Norway.

Here are four surprising facts about Charles Paxton:

ONE: Of the four ostrich researchers, he was the ostriches' sexual favourite.

TWO: It would be misleading to say that he studies ostriches. Paxton no longer works with long-necked, sexually aggressive birds. These days his main research work concerns fish.

THREE: He is a friend of the celebrated and glamorous biologist Olivia Judson, whose book, *Dr Tatiana's Sex Advice to All Creation*, which presents detailed deliciously graphic how-to sex advice from the fictional doctor to a variety of fish, birds, reptiles, mammals, slime moulds, and other species, was all the rage several years ago. Paxton and Judson were students together at the University of Oxford.

FOUR: It is slightly misleading to say, as some people do, that he studies sea monsters. What he studies are reports about sea monsters. Sea monster reports are, for him, part scholarly, part pastime.

You may have noticed hints of a pattern in facts number one, two, and three: sex. Charles Paxton's newest sea monster report, published in the *Archives of Natural History*, continues the pattern. It gives a fresh interpretation to an old sea-monster sighting.

In 1741, a Danish-Norwegian missionary named Hans Egede published what became a famous account of 'a most dreadful monster' that appeared off the coast of Greenland. 'The case is interesting,' the modern scientists write, 'in that Egede had drawn and described a number of large northern whale species in his book so he obviously felt the "dreadful" monster was something different.'

Paxton says that most historians have relied solely on a bad translation of Egede's book. He and his colleagues applied modern biological insights to the case.

Egede's animal had a serpent-like tail that appeared out of the water when the rest of the beast had disappeared. But rather than a tail, Paxton et al. say, this was most likely a penis. They present photographs of well-equipped male whales, and also a drawing from Egede's book, in which we see the sea monster's serpent-like tail. The latter is remarkably similar to what we see in the photographs.

Detail from Egede's monstrous account of 1741 (top); North Atlantic right whale penis of 2001 (bottom). Photograph reproduced by permission of New England Aquarium, Boston, Massachusetts.

The case is not proved definitively, but it should be an inspiration to both biologists and whale-watching tourists.

Paxton, C. G. M., Erik Knatterud, and Sharon L. Hedley (2005). 'Cetaceans, Sex and Sea Serpents: An Analysis of the Egede Accounts of a "Most Dreadful Monster" Seen Off the Coast of Greenland in 1734.' *Archives of Natural History* 32 (1): 1–9.

FOLD WHEN WET (IF NAKED)

Yuri Glebovich Aleyev used an electric winch to tow naked women underwater at speeds of two to four metres per second. Later, his colleagues, when they peered at Aleyev's films and photos, had reason to be upset. What they saw was not what anyone, except maybe Aleyev, was expecting.

Aleyev, who died in 1991, was one of the world's great experts on nekton, which is an obscure word for animals that swim where they wish, rather than merely drifting along. Plankton are not nekton. Fish, dolphins, and people are. Aleyev spent much of his life and ingenuity trying to tease out the secrets of how good-swimming creatures swim so well. The naked women served as stand-ins, so to speak, for wild dolphins.

Aleyev wanted to test something many of his colleagues believed: that dolphins slip so easily through the seas because their skin forms special, undulating folds. Those folds, hypothesis had it, keep the water flowing smoothly – rather than turbulently – past the speeding dolphin.

Others had tried photographing dolphins in action, expecting to capture clear images of mighty, mobile ripples travelling down their bodies. However, film after film failed to show the telltale lines. Thus came Aleyev to the quarrel, and thus, at his invitation, came forty professional swimmers to a pool. Using basic, pithy engineering language (including a mention of the difficult-to-describe-in-words Reynolds number), Aleyev explained that: 'women are similar in body size to average-sized dolphins of the Delphinus type. For women 160–170 centimeters tall swimming with arms stretched forward at a speed of 2.0–4.0 meters/second, the range of Reynolds numbers is about 3.0×10^6 to 9.0×10^6, which is entirely inside that most usual for dolphins... The body surface of the typical woman may be considered to a sufficient approximation hairless, which is characteristic also of dolphins.'

In the early 1970s, Aleyev produced three papers about his experiments. He later summarized them, along with many of his other discoveries, in a book called *Nekton*, written in Russian. An English translation came out in 1977 from a Dutch publishing company with the curious name Dr W. Junk. The volume includes a generous selection of action photos of the women, who are not quite as hairless as advertised, and a few corresponding pictures of dolphins.

The images tell a tale that Aleyev interprets in the accompanying text. Skin ripples do appear, but only when the women (and the dolphins) are in a sharp spurt of acceleration or when they move at the very highest speed. These are not at all 'the result of the contraction of certain trunk and skin muscles'. They are merely passive ripples in the aquatic breeze, akin to wind-furls in a flying flag. And when the skin-folds form, they probably slow down the swimmer, rather than speed her up. Thus did Yuri Aleyev and his underwater camera and his electric winch, assisted by forty skilled swimmers, destroy a biological doctrine of his day.

It was a Fish who told me about Aleyev's experiment – Frank E. Fish, who studies fish and who, when above sea level, often legitimately acts the role of a biology professor at West Chester University in Pennsylvania.

Aleyev, Yuri Glebovich (1977). *Nekton*. The Hague: Dr W. Junk. See 264–78.

ON MONKEY VOMIT (FOR THOSE WHO WANT TO KNOW)

'Researchers have given little consideration to vomiting in nonhuman primates.' Quite so. A new report called 'Vomiting in Wild Bonnet Macaques' begins with that statement, and tries to remedy the deficiency.

Elizabeth Johnson, Eric Hill, and Matthew Cooper published their study in the *International Journal of Primatology*. Johnson is at Oglethorpe University in Atlanta, Georgia. Hill is at Arizona State University, and Cooper at Georgia State University.

They start with a fond look back at the work of earlier vomiting experts. The consensus view, they say, is that vomiting 'is a theoretically complex behavior that to date lacks a comprehensive explanation'.

Johnson, Hill, and Cooper spent time with macaques, carefully noting when each individual animal vomited and whether it then reingested (for that is the technical term) whatever came up. All told, the scientists compiled 'both quantitative and qualitative data on observations of 163 instances of vomiting from 2 groups of bonnet macaques in southern India'. They used this data to 'establish a conservative rate of vomiting in free-ranging macaques'.

The rate is 0.0042 vomits per individual per hour. That's the conservatively high estimate, using data gathered by watching macaques that live in and near a temple on top of Chamundi Hill, a rocky, forested outcrop near Mysore, in Karnataka, India. But it's not the whole story. Another group of macaques lives in the Indira Gandhi Wildlife Sanctuary, Anaimalai Hills, Tamil Nadu. These forest-dwellers vomit at a different rate from their temple cousins: 0.0028 vomits per macaque per hour.

The scientists observed closely and keenly. Here is a typical passage from their report: 'Only 1 adult female in the forest showed interest in another macaque's vomit; she twice smelled the mouth of an adult female. During observations at the temple we saw 20 different individuals show interest in another's vomit on 21 occasions. Ten of the individuals were successful in eating some of it on 11 occasions. Of the individuals that ate or tasted another monkey's vomit, 2 were adult females, 2 were adult males, 3 were juvenile females, and 3 were infants.'

RESULTS

We easily recognized vomiting. The individuals typically made slight heaving motions with the head and shoulders, stomach contractions produced sound, and vomited material was often present on their lips. They manipulated the material in their mouths and cheek pouches, and usually
---
Detail from 'Vomiting in Wild Bonnet Macaques'

The study builds to a thrilling conclusion. The researchers explain what, to them, is a central mystery about vomiting in wild bonnet macaques. Why, they ask, don't the macaques simply vomit and walk away? Why do they immediately 'reingest' the vomit?

Earlier scientists seem not to have noticed this mystery or, if they did notice, to have offered a good explanation.

The key, according to Johnson, Hill, and Cooper, lies in a simple fact. Macaques have spacious pouches built into their cheeks. Johnson et al. apply some logic. 'We suggest', they write, 'that the tendency to hoard food in their cheek pouches explains why they reingested the vomit.'

The study concludes with a modest statement: 'Our data offer insight into a normal, but largely ignored, behavior of cercopithecines.'

Johnson, Elizabeth C., Eric Hill, and Matthew A. Cooper (2007). 'Vomiting in Wild Bonnet Macaques.' *International Journal of Primatology* 28 (1): 245–56.

FOUR

BEHAVE YOURSELF (OR DON'T)

IN BRIEF

'I'M WAITING FOR THE BAND: PROTRACTION AND PROVOCATION AT ROCK CONCERTS'
by Richard Witts (published in *Popular Music*, 2005)

Some of what's in this chapter: Sheep personality profiling • Spaced out on the beach, measurably • Seating in cinemas, handedly (in Bulgaria) • Booing at bigshots • Looking at liars, internationally • Punks and accountants • Clowns of a ministerial turn • Racial cheese profiling • Watching people watch their laundry • Naked in the library • A toilet dilemma • Oscillating, as one will do • Noise-making among the elderly • Chewing delicious food • Chewing distasteful food

CLING BOLDLY, SHEEP

To know – rather than guess – why certain sheep cling to each other while others split off on their own, a person would need to know the size of the group, and also know something about the personalities of the individual sheep. Scientists at the Macaulay Institute in Craigie-buckler, Aberdeen, Scotland, sought this very knowledge when they looked at the loiterings of sheep.

Pablo Michelena, Angela Sibbald, Hans Erhard, and James McLeod (the names of the scientists, not the sheep) wrote up their adventure in a study called 'Effects of Group Size and Personality on Social Foraging: The Distribution of Sheep Across Patches'. It appeared in 2009 in the journal *Behavioral Ecology*.

The four scientists observed the meanderings of fifty-eight female

Scottish Blackface sheep in tasty green fields, under tightly controlled conditions. Before letting the sheep loiter, the researchers gave each of them a personality test, noting which sheep were bold enough to explore objects with novel smells (lavender, mint, thyme, marjoram, garlic, or coffee) and exotic shapes (a baby's rattle, a bottle brush, and various baby's teething rings).

Then they let the sheep loose, in groups of two, four, six, or eight, to graze in grassy arenas. Each arena had some patches of especially desirable (in the scientists' educated opinion) greenery. That extra-yummy fodder, sprung from extra-fertilized soil, and was allowed to grow especially long so as to be extra-noticeable to the sheep.

In the scientists' view, the sheep faced a dilemma: 'In our study, sheep faced a trade-off between maximizing their access to a preferred, but limited, resource and staying together as a group.'

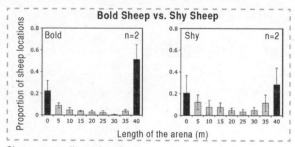

Sheep personality vs. location

More often than not, groups broke apart. And here personality came to the fore, say the scientists: 'bold sheep ... tended to split into subgroups at smaller group sizes than shy sheep'. That was the study's major finding. The scientists discovered that after a split, the new, little groups would often be of equal size.

The idea that each individual sheep might have a uniquely distinctive personality is quite modern, academically speaking. Until recently such individuality in non-human animals had never been documented by scientists, at least not in a way that merited bold, unqualified mention by other scientists.

Michelena et al. write about the newness of their notion. They say: 'Comparative psychologists and behavioral ecologists have recently documented consistent intraspecific differences between individuals in traits such as aggressiveness, activity, exploration, risk taking, fearfulness, and emotional reactivity.'

When they say 'recently' they mean 1998, when a treatise in the *Philosophical Transactions of the Royal Society of London* persuaded many biologists that some pumpkinseed sunfish are shy, and others are bold.

Michelena et al. note that earlier studies did point the way towards these personality insights. A prime example, they say, is the distractingly named 'The Relation Between Dominance and Exploratory Behavior is Context-Dependent in Wild Great Tits', which delighted ornithologists in 2004.

Michelena, Pablo, Angela M. Sibbald, Hans W. Erhard, and James E. McLeod (2009). 'Effects of Group Size and Personality on Social Foraging: The Distribution of Sheep Across Patches.' *Behavioral Ecology* 20 (1): 145–52.

Wilson, D. S. (1998). 'Adaptive Individual Differences within Single Populations.' *Philosophical Transactions of the Royal Society of London* B 353: 199–205.

Dingemanse, N. J., and P. de Goede (2004) 'The Relation Between Dominance and Exploratory Behavior is Context-Dependent in Wild Great Tits.' *Behavioral Ecology* 15: 1023–30.

SPACING AT THE BEACH

Some thirty years ago, beachgoers in three countries found that strangers were coming up to them, asking strange questions. The strangers turned out to be fairly harmless. They were academics, driven by a fierce desire to understand how much space people appropriate for themselves when they plop down on a beach.

Until 1974, only lifeguards and the beachgoers themselves knew the answer. No one in academia had sufficient data to address the question with any degree of authority.

In midsummer of the previous year, Julian Edney and Nancy Jordan-Edney of the University of Arizona had travelled two thousand miles east and spent five days striding up and down a beach. Their subsequent report, called 'Territorial Spacing on a Beach', published in the journal *Sociometry*, was a landmark in the history of studying territorial spacing on beaches.

The Edneys' artfully collected data, after careful crunching and interpretation, told them several things. As groups get bigger, they tend to grab less space per person. Men tend to grab more space than women. And there were nuances that were not so easily interpreted, then or now.

Seven years later, another American, H. W. Smith, at the University of St. Louis, went to Europe, determined to measure the spacing between people on a beach in France, and then on a beach in Germany. Smith succeeded. His report 'Territorial Spacing on a Beach Revisited: A Cross-National Exploration' subsequently appeared in the journal *Social Psychology Quarterly*.

In both Germany and France, Smith found much the same thing that the Edneys had seen in America. And he discovered something more. 'Lone Germans', Smith wrote, 'had more circularly shaped claims than lone French persons.' Also, Germans 'overwhelmingly (99%) tended to structure very rigidly public space by building sand castles around their territories'.

The urge to measure people's personal space has not been confined to beaches. In 1974, Paul Nesbitt of the University of Nevada, Reno, and Girard Steven of the University of California, Santa Barbara, published 'Personal Space and Stimulus Intensity at a Southern California Amusement Park'. They explain how they sent an attractive young woman, or alternatively a man, into the queues for various attractions at an amusement park. 'It was found that subjects immediately behind them in line stood further away when the stimulus persons wore brightly colored clothes than when they wore conservative clothing. Subjects similarly stood further away when the stimulus persons used perfume or after-shave lotion than when they used no scent.'

Recently, Masae Shiyomi, of Ibaraki University in Mito, Japan, performed an Edney-esque set of measurements with cows. Details can be found and enjoyed in her report 'How are Distances Between Individuals of Grazing Cows Explained by a Statistical Model?' This is the sixth in Shiyomi's ongoing and subtle series of cow-spacing reports.

Cows in a pasture, she finds, space themselves differently than do people on a beach. How, exactly, do cows form a crowd? The question drives Shiyomi; statistically minded farmers will want to follow her adventures.

Edney, Julian J., and Nancy L. Jordan-Edney (1974). 'Territorial Spacing on a Beach.' *Sociometry* 37: 92–104.
Smith, H. W. (1981). 'Territorial Spacing on a Beach Revisited: A Cross-National Exploration.' *Social Psychology Quarterly* 44 (2): 132–37.
Nesbitt, Paul D., and Steven, Girard (1974). 'Personal Space and Stimulus Intensity at a Southern California Amusement Park.' *Sociometry* 37 (1): 105–15.
Shiyomi, Masae (2004). 'How are Distances Between Individuals of Grazing Cows Explained by a Statistical Model?' *Ecological Modeling* 172: 87–94.

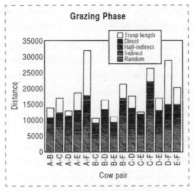

Distance between two cows during grazing phase

TAKE A SEAT IN BULGARIA

When people walk into a cinema, where do they choose to sit? The question has vexed several brain researchers.

The topic arose in Bulgaria. Bulgarian cinema receives less global attention than its counterparts in other developed countries. Bulgarian cinema audiences receive correspondingly little scrutiny. This attention deficit was addressed, slightly, in the year 2000, when George B. Karev of the Bulgarian Academy of Sciences conducted his study 'Cinema Seating in Right, Mixed and Left Handers'.

At the time, Karev was best known for his 1993 report 'Arm Folding, Hand Clasping and Dermatoglyphic Asymmetry in Bulgarians'. The cinema seating study, dealing as it does in questions of left versus right, in some respects builds on the earlier work.

Karev made some diagrams showing the seat locations in five different cinemas. He blocked off the seats in the middle, and asked people to tell him which of the open seats they would select. Most chose seats on the right side. This was especially true among people who, in answer to another question, said they were right-handed.

Why this general preference for the right side? Most probably, Karev says, it's because: (a) films pack an emotional wallop; (b) one side of the brain is better at handling emotions; and (c) experienced film-goers learn to sit where that side of their brain will have the best vantage point.

The response of the scientific community was immediate, if minuscule. Professor Sergio Della Sala of the University of Aberdeen suggested that 'one possible way to find out if Karev is correct would be to ask people to sit in a room exempt from any emotional content – example a large waiting room, a lecture theatre, even possibly the House of Lords?' Della Sala made this comment in the form of a press release. The press release announced two things: that Karev's study had just been published; and that Della Sala was the new editor of the journal that published it. The journal is called *Cortex*.

That was about the extent of the scientific community's reaction to the Karev experiment, at least publicly, until 2006. In that year, a German research quartet took the stage.

Peter Weyers and colleagues at Bavarian Julius-Maximilians University repeated Karev's experiment, but with some twists. The original cinema diagrams showed the film screen at the top of the page. But here, some diagrams showed the screen at the bottom of the page, or on one side. Looking at these diagrams, people had no real preference for sitting with the screen to their left or to their right.

The Germans published a report in the journal *Laterality*. There could be many reasons, they said, why the Bulgarians opted for the right. Top of the list: the odd fact that most people habitually turn to the right when entering a room.

That's how things stand, for now, on the mental and cinematic significance of choosing sides in Bulgaria or elsewhere.

The journal *Laterality*, by the way, is edited by Chris McManus of University College London. Professor McManus was awarded the 2002 Ig Nobel Prize in biology for the short treatise 'Scrotal Asymmetry in Man and in Ancient Sculpture', which he wrote soon after graduating from medical school. The journal *Nature* published the article in 1976, and featured it on their front cover.

Karev, George B. (2000). 'Cinema Seating in Right, Mixed and Left Handers.' *Cortex* 36 (5): 747–52.
— (1993). 'Arm Folding, Hand Clasping and Dermatoglyphic Asymmetry in Bulgarians.' *Anthropologischer Anzeiger* 51(1): 69–76.
Weyers, Peter, Annette Milnik, Clarissa Müller, and Paul Pauli (2006). 'How to Choose a Seat in Theatres: Always Sit on the Right Side?' *Laterality* 11 (2): 181–93.

BOOS ACT AS BOOZE ON THE POWER-HUNGRY

People with a tremendous drive for power sometimes encounter obstacles. An experiment measured what happened when power-driven people gave speeches to an audience that responded with blatant, deliberate acts of boredom.

The researchers, Eugene Fodor and David Wick of Clarkson University in Potsdam, New York, wrote up the details in a blandly titled monograph, 'Need for Power and Affective Response to Negative Audience Reaction to an Extemporaneous Speech'.

Fodor and Wick found some power-seekers and, for comparison, some power-avoiders. They used a standard psychological method to discriminate these people from those sorts who are merely indifferent to or accepting of power. They asked each volunteer to write little stories about a set of pictures. The pictures showed '(1) seven men around a table, (2) man with cigarette behind woman, (3) architect at desk, (4) two women in lab coats in laboratory, (5) ship captain, and (6) trapeze artists'. The volunteers' stories revealed, at least in theory, who unconsciously craved power and who did not.

Having selected their test subjects, Fodor and Wick asked each to give a three-minute persuasive speech to an audience. A wee, special audience it was – one woman and one man specially trained and rehearsed for the occasion.

Fodor and Wick 'predicted that power-motivated participants would exhibit higher levels of electromyographic activity in the brow supercilii when confronted by a negative audience reaction to their speech'. In this, too, they relied on an established method, trusting that the electrical activity level in the forehead-frowning muscles would reliably indicate a person's anxiety level.

For some speechgivers, the audience showed interest. But for others, not: 'Fifteen seconds into the speech, the young woman crossed her legs and began looking at her hands. The young man began to shift in his chair. The woman continued looking around. The man looked at his watch, then briefly gaped out the window. At approximately 1 min into the speech, the actors looked at each other and raised their eyebrows. They then looked back at the participant delivering the speech. Both continued to shift their gaze to their hands or the floor, rarely looking at the participant. Approximately 2-and-1/2 min into the speech, near the end, the woman gave off a visible sigh. The actors continued to look around for the remaining 30s. The man, for his part, twiddled his thumbs a lot, looked at the clock a few times, yawned at specific junctures.'

The results of the experiment: under this kind of duress, the power-hungry persons, compared to the non-power-hungry individuals, had noticeably greater eyebrow-furrowing-muscle electrical activity.

Fodor and Wick end their report with an eyebrow/anxiety-raising cautionary note for anyone who aspires to leadership. They specifically mention politicians and labour-management negotiators: 'The findings ... suggest that certain occupations may pose repeated exposures to stress of a kind that can threaten cardiovascular health for persons high in power motivation.'

Fodor, Eugene M., and David P. Wick (2009). 'Need for Power and Affective Response to Negative Audience Reaction to an Extemporaneous Speech.' *Journal of Research in Personality* 43: 721–26.

LIAR, LIAR

In 2006, a group called the Global Deception Research Team published a report called 'A World of Lies'. It appeared in the *Journal of Cross-Cultural Psychology*.

The team is big. It has ninety-one members, spread all around the world. Their stated goal: 'studying stereotypes about liars'.

They ask someone, 'How can you tell when people are lying?', then follow this up with ten simple multiple-choice questions about liars:

- When people are lying, they act ... calm, nervous, or neither calm nor nervous?
- When people are lying, they act ... silly, serious, or neither silly nor serious?
- When people are lying, their stories are ... more consistent than usual, less consistent than usual, or neither?
- When people are lying, their stories are ... longer than usual, shorter than usual, or neither?
- Before answering questions, people who are lying pause ... longer than usual, shorter than usual, or neither?

- When people are lying, they stutter ... more than usual, less than usual, or neither?
- When people are lying, they shift their posture ... more than usual, less than usual, or neither?
- When people are lying, they look at the other person's eyes ... more than usual, less than usual, or neither?
- When people are lying, they touch and scratch themselves ... more than usual, less than usual, or neither?
- When people are lying, they use hand gestures ... more than usual, less than usual, or neither?

They asked these questions of people in sixty-two different countries: China, Colombia, Croatia, Cyprus, the Czech Republic, the Dominican Republic, Egypt, Estonia, Finland, France, Georgia, Germany, Ghana, Greece, India, Indonesia, Iran, Ireland, Israel, Italy, Japan, Jordan, Kenya, Korea, Kuwait, Lithuania, Malaysia, Malta, Mauritius, Mexico, Micronesia, Moldova, Morocco, Nepal, the Netherlands, New Zealand, Norway, Pakistan, Paraguay, Peru, Philippines, Poland, Portugal, Romania, Russia, Samoa, Serbia, Slovakia, Slovenia, South Africa, Spain, Sri Lanka, Swaziland, Sweden, Switzerland, Taiwan, Togo, Trinidad and Tobago, Turkey, the United Arab Emirates, the UK, and the US.

The Global Deception Research Team compiled and analysed the answers. They distilled it down to this: '[There are] common stereotypes about the liar, and these should not be ignored. Liars shift their posture, they touch and scratch themselves, liars are nervous, and their speech is flawed. These beliefs are common across the globe. Yet in prevalence, these stereotypes are dwarfed by the most common belief about liars: "they can't look you in the eye".' That is their great discovery. And it accords with previous discoveries by other researchers.

The team prepared for its work by studying thirty-two earlier studies about lying. A 1981 survey of Americans, they say, found the widespread belief that 'liars avert gaze, touch themselves, move their feet and legs, shift their posture, shrug, and speak quickly'. A 1996 survey of Britons revealed the general opinion that 'liars reduce eye contact, turn away, blink, and pause while giving inconsistent, implausible stories'.

Of these and other nations' beliefs, the Global Deception Research Team says 'These beliefs are probably inaccurate.' It is well established, they say, that people show little ability to detect when somebody is lying.

The Global Deception Research Team did not ask whether the people who answered their survey were lying. The reader may presume that the researchers presume that, when people answer surveys, they tell the truth.

The Global Deception Research Team (2006). 'A World of Lies.' *Journal of Cross-Cultural Psychology* 37 (1): 60–74.

NORMAN, THE PUNK OR THE ACCOUNTANT

The finding about punks and accountants came in two parts. The finders, University of Exeter psychologists Louise Pendry and Rachael Carrick, published their study in the *European Journal of Social Psychology*. In essence, their research is not really about punks and accountants – rather, it's about conformity.

Pendry and Carrick's first insight, though small and unshocking, is technically unprecedented in the annals of psychology. They got it by recruiting run-of-the-mill, non-punk, non-accountant individuals, and asking them sly questions. The answers, Pendry and Carrick say, 'revealed that a category strongly associated with non-conformity was that of punks; whereas for conformity, a popular group was that of accountants'.

Pendry and Carrick's second, greater insight came from an experiment. In this, too, the test subjects were neither accountants nor punk rockers. They can be thought of, in a purely academic sense, as innocent dupes.

The basic idea, with each of the dupes, was to show them a picture, then see how that picture had affected them. Would the innocent dupe be more – or less – willing to conform with other people's opinions?

Pendry and Carrick describe the set-up tersely: 'Participants were given a photograph of either an accountant or a punk and instructed to study it carefully for a few moments. The accountant photo depicted a man with neat appearance, wearing a suit, with short hair, and glasses. The punk photo showed a young man with spiky hair, and torn clothing covered in graffiti.'

For the sake of clarity, Pendry and Carrick embedded words within each photograph: either 'Norman, who is an accountant' or 'Norman, who is a punk rocker'.

The dupe being thus properly set up, he or she was then crammed into a room with three non-dupes and an authority figure. The authority figure played a tape recording full of beeps, first asking everyone to (1) pay attention and (2) carefully count the beeps. After the playing of the beep-filled tape came the moment of truth ... or the moment of conformity.

The authority figure asked each of the confederates how many beeps they'd heard. Each of these co-conspirers gave a pre-arranged and wrong total.

Now, at long last, the innocent dupe had to speak up. How many beeps had she or he heard?

The innocent dupes who had seen the photo of an accountant fudged their answer. They acquiesced to what everyone else said. The dupes who had looked at a punk rocker did not.

Like many studies, this one builds on an existing foundation. Pendry and Carrick acknowledge owing much to a 1996 New York University study about innocent dupes who were shown a list of words about elderly people. The words included old, lonely, grey, retired, wrinkle, ancient, and cautious. The scientists, armed with a stopwatch, discovered that dupes who had seen those words walked away more slowly than dupes who had not.

Pendry, Louise, and Rachael Carrick (2001). 'Doing What the Mob Do: Priming Effects on Conformity.' *European Journal of Social Psychology* 31: 83–92.
Bargh, John, Mark Chen, and Lara Burrows (1996). 'Automaticity of Social Behavior: Direct Effects of Trait Construct and Stereotype Activation on Action.' *Journal of Personality and Social Psychology* 71 (2): 230–44.

MINISTRY OF CLOWNS

Clown Doctors: Shaman Healers of Western Medicine

The Big Apple Circus Clown Care Unit, which entertains children in New York City hospitals, is compared with non-Western healers, especially shamans. There is not only superficial resemblance—weird costumes, music, sleight of hand, puppet/spirit helpers, and ventriloquism—but also

Angelika Richter and Lori Zonner have a funny way of captivating readers. In a study called 'Clowning: An Opportunity for Ministry' they write: 'Experiences over five years interacting with patients as the clown Jingles and the experiment and experience of one afternoon as the clown Hairie in a hospital led the authors to reflect on the deeper meaning of clowns ... Before sharing further experiences with clowning in ministry, and telling about one afternoon when Jingles and Hairie were on their way through the hospital, let us first describe a common meaning of clowning.'

Richter, a chaplain and minister at Philipps University in Marburg, Germany, and her colleague Zonner published their monograph in 1996 in the *Journal of Religion and Health*.

Clowning, as commonly recognized, is for them just a beginning. Richter and Zonner explain that 'the clown is recognized universally as a symbol of happiness and creates smiles and laughter. The clown ministry, however, is not just entertainment, nor is it preaching in a costume.'

Looking beyond that research, one sees that clowning ministry is often confined to hospitals, but not to any one country. In Scotland, Olive Fleming Drane of Aberdeenshire proudly administers the yuks. In England, Roly Bain of Bristol is the most prominent of this variety of spiritual clown. The US is bursting with clowns of a ministerial turn.

For anyone wishing to be initiated, resources abound.

Janet Litherland's book *The Clown Ministry Handbook*, published in 1982, offers something of a one-stop education. The table of contents lays out the basics: 'An overview of the activities of clowns throughout history'; 'The "where" and "how" of clown ministry'; 'How to entertain an audience by making a wide variety of objects from balloons'; and more. The final chapter crowns it: 'Eleven clown ministers tell how they came to be clowns for Christ'.

A website called ClownMinistry.com gives clown ministry info and instruction, and sells clown ministry paraphernalia ranging from *The Clown Ministry Handbook* to Three Stooges golf ornaments. The day I visited, the site featured a sponsored link to MyGunSpot, a 'social networking site for gun owners'.

However, not everyone loves a clown, even a worshipful clown. And sometimes, clownish optimism meets donnish discouragement.

Linda Miller Van Blerkom, of Drew University in New Jersey, published a study in *Medical Anthropology Quarterly*, where she cautioned that: 'small children are frequently afraid of clowns, whose bizarre appearance suggests the dangers of the unknown and uncanny, and whose performances dramatize common childhood fears'.

To clown-lovers, Miller Van Blerkom's work may sound flat, lifeless, sterile. But the Economic and Social Research Council warned in 2007 that even two-dimensional artwork of clowns, affixed to a wall of a hospital, can be problematic. Citing research performed by Penny Curtis of the University of Sheffield (and which it sponsored), the council issued an alert to hospitals, in 2007, with the headline 'Children's Wards – Don't Send in the Clowns'. The most chilling detail: 'All children disliked the use of clowns in the décor, with even the oldest children seeing them as scary.'

Richter, Angelika, and Lori A. Zonner (1996). 'Clowning: An Opportunity for Ministry.' *Journal of Religion and Health* 35 (2): 141–48.

Miller Van Blerkom, Linda (1995). 'Clown Doctors: Shaman Healers of Western Medicine.' *Medical Anthropology Quarterly* 9 (4): 462–75.

AN IMPROBABLE INNOVATION

'ODOR GENERATION ALARM AND METHOD FOR INFORMING UNUSUAL SITUATION'

a/k/a a wasabi-fume-emitting alarm by Makoto Imai, Naoki Urushihata, Hideki Tanemura, Yukinobu Tajima, Hideaki Goto, Koichiro Mizoguchi, and Junichi Murakami (US Patent application no 2010/0308995 A1, filed 2009 and honoured with the 2011 Ig Nobel Prize in chemistry)

THE CHEESE FILES

Because race is an uncomfortable topic for many people, certain questions simply do not get discussed. It is now nearly thirty years since the publication of Beth A. Scanlon's blockbuster report 'Race Differences in Selection of Cheese Color'. In all that time, the report has received nary a mention in public forums.

I have found no reference to Scanlon's report in any political speech, anywhere. This is not surprising. No skilled politician likes to venture near a potentially divisive subject on which public sentiment is still unclear.

Scholars, on the other hand, sometimes love to stake out an early position on a controversial issue. It's a simple way to make a name for oneself in the professional community. But the academic world, too, has been virtually silent on the question of race differences in selection of cheese colour.

Only one other published academic paper pays any attention to the Scanlon race/cheese-colour report. And the paper, published in the *Journal of Marketing Theory and Practice*, does it glancingly, in a curious sentence that begins: 'In some cases, subjects are encumbered with: red eye-goggles (DuBose et al. 1980), blindfolds (Hyman 1983; Scanlon 1985), red lights (Hall 1958; DuBose et al. 1980) or red glass (Duncker 1939) to mask color; funnels and jugs for spitting (Looy, Callaghan & Weingarten 1992)'.

The emphasis in that little mention, obviously, is on blindfolds. But blindfolds are a mere detail of Beth Scanlon's experiment. Her quest – to explore race differences in selection of cheese colour – gets overlooked.

The Scanlon report itself is brief – just one page long. And it is blunt. 'White and yellow American cheese was presented to 155 individuals from three ethnic groups', Scanlon writes. One group is black, one white, the other Hispanic. 'In a supermarket, a display table was set with two plates of American cheese, one yellow, one white. As the individuals selected a piece of cheese, the grouping and the color of the chosen cheese was recorded.'

Scanlon also offered the cheese to an extra, so-called 'control' group of people, each of whom was blindfolded. The blindfolded cheese-samplers, she says, 'reported no significant difference in flavor of the cheeses'.

The overall results of the experiment? Scanlon concludes that 'the preferences for one of two colors of American cheese are dissimilar for different races of respondents'.

As far as I could determine, this is the only research report Beth A. Scanlon ever published. She was – though is no longer – based at Central Connecticut State University. What was her intent in exploring race differences in selection of cheese colour? Why did no one pick up on and continue her line of research? Why did Scanlon herself drop the question, and what has she done with her time instead? These remain mysteries.

Scanlon, Beth A. (1985). 'Race Differences in Selection of Cheese Color.' *Perceptual and Motor Skills* 61 (1): 314.

Garber, Lawrence L., Jr, Eva M. Hyatt, and Richard G. Starr, Jr (2000). 'Placing Food Color Experimentation into a Valid Consumer Context.' *Journal of Marketing Theory and Practice* 8: 59–72.

IN BRIEF

'SCRATCH AND SNIFF: THE DYNAMIC DUO'
by W. Z. Stitt and A. Goldsmith (published in the *Archives of Dermatology*, 1995)

LAUNDRY MARKS

In a laundromat, how do people behave? Scholars mostly avoided the question until the early 1980s, when Regina Kenen, an assistant professor of sociology at Trenton State College in New Jersey, became the first sociologist to camp out in a middle-class laundry and take detailed notes.

Scholars as a group have a mixed reputation about personal cleanliness in general, and clothes-washing in particular. Kenen subtly addresses this early in her report, which is entitled 'Soapsuds, Space, and Sociability: A Participant Observation of the Laundromat'. She gathered her data, she tells us, in 'the San Francisco Bay area laundromat that I used regularly'.

Kenen sketches her fellow clothes-washers for us. 'The apparel they wear is very informal. Occasionally, some women come in heavily made up, wearing stiletto heels, stockings, and dressy clothes. They stick out as oddities; even more rarely, men wear suits.'

Then it's down to business: keen descriptions of these people's actions and interactions. For the lay reader, there are insights aplenty.

Customers 'glance around to see where there are empty washing machines but do not ordinarily look at individuals directly ... If the laundromat is fairly empty and they have the choice, they often leave an empty machine between theirs and adjacent users'.

The launderers don't interact much. There are, however, some key exceptions. Those who come to the laundromat together or meet a friend there 'converse, laugh, and touch while engaged in the tasks and there is a sense of mutuality and involvement with each other that clearly signals that they are a unit and not interested in further interaction with others.' Lone individuals 'maintain more solemn facial expressions than do couples, and they do not talk to strangers except in a purely functional way, e.g., to say "excuse me" when they are trying to move their cart full of wet clothes to the dryer or to say to someone, "you dropped something".'

Some customers leave, and return after their clothes are done. Others stay the whole time. The hangers-on engage in a variety of behaviours.

Some sit and read. Kenen categorizes them into four distinct types. The 'desultory' reader 'merely flips pages of a magazine or newspaper'. The 'interested' reader reads newspapers or magazines 'with seeming intent and concentration'. The 'involved' reader brings his or her own books 'and is completely oblivious of the surroundings'. The 'instrumental' reader reads 'textbooks and other assigned materials'.

Kenen also saw people eating. She concluded that 'much snacking occurs in the laundromat and seems to serve some of the same purposes as it does in the rest of the society'.

Kenen later went to a laundry in a poor, Latino neighbourhood. There, customers socialized more than did the customers in the other laundromat. This influenced the study's ultimate conclusion: 'Laundromat behaviors appear to be more influenced by the larger sociocultural context in which they are enmeshed.'

This remains, all these years later, sociology's most comprehensive statement of how people behave in a laundromat.

Kenen, Regina (1982). 'Soapsuds, Space, and Sociability: A Participant Observation of the Laundromat.' *Journal of Contemporary Ethnography* 11 (2): 163–83.

THE NUDIST RESEARCH LIBRARY

The American Nudist Research Library has a fairly simple motto: 'Dedicated to preserving nudist history with a comprehensive archive of nudist material'. Like all specialist libraries, it operates with a limited budget. Thus, the library covers only what it needs to.

The institution marked its twenty-fifth anniversary in 2004. The celebratory material explained that 'the Library was established in 1979 to preserve the history of the social nudist movement in North America and throughout the world. It is a repository of material rather than a circulating library. Visitors may read or view most of the collection as long as they are in the Library.'

The facility is in Kissimmee, Florida, on the grounds of Cypress Cove Nudist Resort, just a few miles from Disney World. Visitors are welcomed, whether or not they come equipped with clothing.

A library is a good place to conduct research. This particular library may be a good place to settle an ever-so-slight controversy in the field of cognitive science. Cognitive scientists, some of them, want to know how looking at nude bodies can affect a person's memory.

Dr Stephen R. Schmidt, a professor of psychology at Middle Tennessee State University, tried to settle the question by showing nude photographs to a group of volunteers. He conducted a series of experiments, which he subsequently described in a report called 'Outstanding Memories: The Positive and Negative Effects of Nudes on Memory'.

Schmidt exposed his volunteers to carefully selected photographs, which he presented in various orders and paced at different time intervals. Here is a partial list of the photos: woman pumping gas; man climbing a mountain; woman sitting at a window reading a newspaper; man stacking wood; woman playing a cello. Some – but not all – of the men were nude. Ditto for the women.

This was a sophisticated follow-up to much earlier experiments that were done by psychologists Douglas Detterman and Norman Ellis. Detterman and Ellis embedded a photo of male and female nudes, which they obtained from an issue of *Sunbathing* magazine, into a series of black-and-white line drawings of common objects, and then showed the lot of them to volunteers. The result: 'Not surprisingly, memory for the nudes was much better than memory for [other items] – approaching 100% correct. However, the presence of the nudes caused amnesia, in that memory for items immediately preceding and following the nudes was poor.'

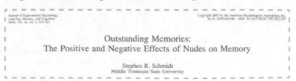

The point of this research? To tease out the subtle nature of why some memories are retained and others forgotten. Why nudes? Because, says Schmidt, 'nudes (rather than other emotional stimuli) seem to provide reliably strong effects'.

Live nudes would seem to provide more reliably strong effects than one would get from photographs of nudes. The American Nudist Research Library has nudes of both varieties, a bounty that should be of interest to scientists.

And it may be instructive to librarians elsewhere who lament that people don't visit libraries the way they used to.

Schmidt, Stephen R. (2002). 'Outstanding Memories: The Positive and Negative Effects of Nudes on Memory.' *Journal of Experimental Psychology: Learning, Memory and Cognition* 28 (2): 353–61.

CHOOSING WHERE TO GO

Where do people go? Though it's a simple question, scholars disagree about where people choose to go to the toilet. What specifically concerns these scholars is a small aspect of the larger puzzle. When you give someone a choice of several, say four, toilet stalls arranged all in a row, which stall do people choose to use?

In the past forty years, there have been two major experimental studies on this topic. The results of the one directly contradict those of the other. The first, in Antarctica, discovered that people prefer the stalls at the ends. The other, in California, found that people prefer the middle stalls.

Figure: 'Comparison of amount of faeces accumulated under the lavatory seats in a lavatory at the Antarctic'

The two experiments were done under greatly differing conditions, so there is plenty of room for argument about what it all means.

Dr H. Hachisuga, a physiologist, spent a winter at Japan's Syowa research station in Antarctica. For reasons that are now obscure, he recorded the amount of faeces that accumulated and froze under each of the base's four adjoining outdoor toilet stalls. Hachigusa used those measurements to estimate the 'frequency of utilization' of each seat. As Hachisuga watched the data mount up, he saw evidence that the end stalls enjoyed considerably higher usage than the middle stalls. He attributed this to what he called the psychological influence of corner preference.

Hachisuga presented a summary of his work in 1972, at a symposium on Antarctic medical studies, held under the auspices of the Japanese Society of Biometeorology. The presentation was later published in the medical journal *Igaku No Ayumi*. It includes a cutaway-view drawing of the four stalls. The two centre stalls are vacant. In each of the end stalls, a seated man labours stoically at his task. Below each of the stalls, a chamber contains a pile of data, the pile size indicative of the stall's popularity.

Professor Nicholas Christenfeld of the University of California, San Diego, monitored four stalls in a public restroom at a California beach. He had the custodian count how many toilet-paper rolls were replaced in each stall over a ten-week period. The results: if toilet-paper consumption translates accurately into stall usage, the middle stalls were used half-again as often as the outer stalls. Christenfeld's terse explanation: people 'reliably prefer the middle ones and avoid the extremes'.

Christenfeld did his California toilet monitorings more than two decades after Hachisuga conducted the Antarctic output experiment. Yet Christenfeld, in his published study, makes no mention of the earlier research – quite possibly he was unaware of Hachisuga's body of work.

This happens often in science, just as it does in other fields of human endeavour. Some bold pioneer steps into little-known territory, ignorant that his is not the very first visit. The traces of these intellectual expeditions, deposited over many years in layers upon the ground, form a sort of mental compost. It sits, ripening, for future scholars to uncover.

Christenfeld, Nicholas (1995). 'Choices from Identical Options.' *Psychological Science* 6 (1): 50–55.
Yoshimura, H. (1973). 'Review of Medical Researches at the Japanese Station (Syowa Base) in the Antarctic.' In O. G. Edholm and E. K. E. Gunderson (eds). *Polar Human Biology*. London: Heinemann, 54–65.

OSCILLATING OVER THIS AND THAT

Everyone oscillates, one way and another. We vibrate, we hum, we bounce. We have our ups and downs. Some of this oscillation attracts the attention of researchers named Tainsh.

In 1972, Michael A. Tainsh published a monograph called 'Oscillation of Human Performance as a Personality Measure', in the journal *Perceptual and Motor Skills*. Tainsh was then based at the University of Aston, an institution whose very name oscillates. In its current phase, the name is Aston University.

Tainsh wrote to me about his monograph, saying it was inspired by a forty-five-year-old book: 'May I suggest that you read the work of Spearman and his seminal work *The Abilities of Man*, written, if I recall correctly, in 1927. The concept of oscillation is described fully in the chapter of the same name.'

I followed his suggestion. Spearman – Charles H. Spearman, professor of philosophy of mind at the University of London – indeed rolled out the concept of human oscillation. Spearman then issued a rousing call: 'Finally, there is the great task of determining how this tendency to oscillate correlates with sex, race, social stratum, parentage, and above all with vocational success.' Tainsh is one of the few who answered that call.

Tainsh explains that his 1972 paper 'is very brief and difficult to decipher unless you understand the background which was three years of PhD work based in UK and USA ... The purpose of the work was to describe the "fits and starts" of human performance in terms of wave functions as many engineers would find quite normal if they were examining linear systems. This was quite new to psychologists and my PhD was well received.'

A wave function is a mathematical description, a formula, of something that vibrates. Scientists who study physics labour to write wave functions that correspond to certain behaviours of light, or of a subatomic particle, or of other idealized physical entities. Einstein, Schrödinger, and other modern physicists never managed to devise a wave function to describe a person. But maybe they never tried.

In 1975, Tainsh and a colleague published a study, perhaps distantly related to the oscillation work, called 'The Influence of Travelling on Decision-Making', in which they 'concluded that there was a reduction in the travellers' capacities to make logical decisions following a

100-mile bus journey'. Two years later, Tainsh published a solo study with almost the same title (he removed the word 'The' from the beginning). Though less than a page in length, the later report packs in a tremendous amount of information before reaching its final sentence, which reads: 'Apparently the influence of long distance travelling on logical thinking may be significant but quite small.'

Another, different Tainsh investigated another, different kind of human vibration, in 1988. Susan M. M. Tainsh of the University of Toronto and four colleagues published 'Noise-Making Amongst the Elderly in Long Term Care' in *The Gerontologist*. The journal explains that 'about 30% of residents presented noise-making behavior'.

OSCILLATION OF HUMAN PERFORMANCE AS A PERSONALITY MEASURE

MICHAEL A. TAINSH
University of Aston in Birmingham

Summary.—The relationship between the frequency of the periodic characteristics of individual behaviour and Spearman's (1927) concept of oscillation is discussed in terms of Eysenck's concept of neuroticism. It is shown that the

The study's centrepiece is 'a typology of noise-making' that identifies, once and maybe for all, the six categories of elderly noise-making. These are: 'purposeless and perseverative noise-making'; 'noise-making in response to the environment'; 'noise-making to elicit a response from the environment'; '"chatterbox" noise-making'; 'noise-making in the context of deafness'; and the exhaustive 'other noise-making'.

Tainsh, Michael A. (1972). 'Oscillation of Human Performance as a Personality Measure.' *Perceptual and Motor Skills* 35 (2): 677–78.
— and G. Winzar (1975). 'The Influence of Travelling on Decision-Making.' *Ergonomics* 18 (4). 427–34.
— (1977). 'Influence of Travelling on Decision-Making.' *Perceptual and Motor Skills* 44 (3). 1106.
Spearman, Charles (1927). *The Abilities of Man*. London: Macmillan and Co.
Ryan, David Patrick, Susan M. M. Tainsh, Vita Kolodny, Bonnie L. Lendrum, and Rory H. Fisher (1988). 'Noise-Making Amongst the Elderly in Long Term Care.' *Gerontologist* 28 (3): 369–71.

CALL FOR INVESTIGATORS

The Oscillating Humans Project, announced here, is searching for a living specimen – an exemplar – of an oscillating human.

Definition: For purposes of the project, an Oscillating Human is someone who consistently, repeatedly, over many years, expresses opinions directly opposite to opinions he or she expressed earlier, always ignoring and/or denying the existence of copious, easily found, clear documentation of the earlier opinions.

Purpose: The exemplary person, once identified, will serve as an example for teachers to use in logic classes. To minimize the chance of lawsuits, the exemplar must be a 'public person', for whom (as stated above) there is copious, easily found, clear documentation of years and years of oscillation.

If you know of an outstanding specimen, please provide:

1.   *The name and a twenty-word biographical sketch of the person.*
2.   *Several Internet URLs pointing to clear, unarguable documentation.*

Send to marca@improbable.com with the subject line:
OSCILLATING HUMANS PROJECT

CHEWING ON KNOWLEDGE

When guests come to dinner, a question may arise: 'Do people chew delicious food faster than they chew distasteful food?' The answer seems to be yes, according to an experiment performed by the team of France Bellisle, Bernard Guy-Grand, and J. Le Magnen at Hotel-Dieu Hospital in Paris. Bellisle, Guy-Grand, and Le Magnen published their mastication report in the journal *Neuroscience and Biobehavioral Reviews*.

It's worth noting that Bellisle made waves in 2001, when she and collaborator Anne-Marie Daliz reported, as part of a larger study, that women who eat lunch while listening to a recorded detective story take in more food than women who don't.

The technical details of the Bellisle, Guy-Grand, and Le Magnen study are worth, as they say, chewing over: 'Cocktail size (3 square centimeters) open sandwiches were served in one of five different flavors. An oscillographic recording of chewing and swallowing showed that chewing activity varied with the palatability and variety of foods. Chewing time was shorter and fewer chews were observed as palatability increased. Swallowing did not change as a function of stimulus flavor. Pause duration between two successive food pieces became shorter as palatability increased. The effects of sensory factors were most evident at the beginning of meals and decreased until the end of meals.'

Let me partially digest that passage, and then regurgitate it in plain language. The scientists make three points:

1) people chew delicious food more quickly than they chew horrible food;
2) people race to put delicious food in their mouths, but with horrible food they hesitate; and
3) people enjoy a meal more when they are hungry than when they are full.

These are good things to know – and we now know them scientifically. But that is not all we know. And edograms figured into our knowledge.

An edogram is a graph with two wavy lines: one line zigs every time a person chews, the other line zags every time the person swallows. In doing their study, the three scientists learned that ordinary people can become highly accustomed to wearing the ungainly equipment used in making the edogram. Here is a condensed passage from the official report: 'During test meals, the strain gauge was placed on the subject's cheek. A small balloon, filled with water, was maintained on the subject's throat by an adjustable elastic collar. The subjects did not report any discomfort from the apparatus. One subject even fell asleep briefly during a meal, with her head resting on the table.'

For hosts and hostesses who obsess about their housekeeping rather than their cooking, this is good news. If you have tasty enough food – or in a pinch, if you simply have enough food – your guests will be able to ignore any distractions. Unless you read them a detective story while they are eating.

Bellisle, France, B. Guy-Grand, and J. Le Magnen (2000). 'Chewing and Swallowing as Indices of the Stimulation to Eat During Meals in Humans: Effects Revealed by the Edogram Method and Video Recordings.' *Neuroscience and Biobehavioral Reviews* 24 (2): 223–28.

Bellisle, France, and Anne-Marie Dalix (2001). 'Cognitive Restraint Can Be Offset by Distraction, Leading to Increased Meal Intake in Women.' *American Journal of Clinical Nutrition* 74 (2): 197–200.

FIVE

EAT, THINK, AND BE MERRY

MAY WE RECOMMEND

'EFFECT OF ALE, GARLIC, AND SOURED CREAM ON THE APPETITE OF LEECHES'

by Anders Barheim and Hogne Sandvik (published in *BMJ*, 1994, and honoured with the 1996 Ig Nobel Prize in biology)

Some of what's in this chapter: Gut rumbling for shrinks • The tasting of the shrew • Taste-testing water, with rats • Eating eggs, eggs, eggs, and then some • How and why to explode meat • Tasty pet food • The Attitudes to Chocolate Questionnaire • Wondering about whisky and candles • Standard glops of food • Teabags • The frailty of bunnies

YOUR GUT SAYS ...

Some psychoanalysts can find meaning in the most ordinary-seeming bits of your life. Some discern it even in your intestinal rumblings. There's a technical name for those digestive sounds: *borborygmi*. Several published studies tell how to interpret people's gut feelings – how to translate those borborygmi into common everyday words.

In 1984, Christian Müller of Hôpital de Cery in Prilly, Switzerland, published a report called 'New Observations on Body Organ Language' in the journal *Psychotherapy and Psychosomics*. Müller paraphrases a 1918 essay, by someone named Willener, which 'concludes that the phenomenon generally known as borborygmi must be regarded as cryptogrammatically encoded body signals that could be interpreted with the help of [special] apparatus'. Müller laments that Willener's 'attempts to follow up on his theory were thwarted by the defects of recording techniques at that time'.

Happily, Müller himself had access to later, better equipment. 'We have been trying at our clinic since 1980', he writes, 'to combine electromesenterography with Spindel's alamograph, and in addition to use digital transformation for a quantitative analysis of the curves via computer.'

Müller reveals his greatest interpretive triumph: 'The presence of a negative transference situation was not difficult to deduce from the following sequence: "Ro ... Pi ... le ... me ... lo ...". The following translation is certainly an appropriate rendering: "Rotten pig. Leave me alone."'

This lovely piece of deadpan, intentional nonsense, I am told, was swallowed whole by some readers, and perhaps also some journal editors.

A few years later, Guy Da Silva, a Montreal psychoanalyst, published several apparently quite serious papers about the psychoanalytical significance of borborygmi.

The most accessible (in my view, anyway) is his 'Borborygmi as Markers of Psychic Work During the Analytic Session. A Contribution to Freud's Experience of Satisfaction and to Bion's Idea About the Digestive Model for the Thinking Apparatus'. This professionally dense monograph appeared in a 1990 issue of the *International Journal of Psychoanalysis*. Freud is Sigmund Freud, the psychoanalysis pioneer who lived in Vienna, Austria. Bion is Wilfred Ruprecht Bion, director of the London Clinic of Psychoanalysis in the 1950s, and later president of the British Psychoanalytical Society.

Guy Da Silva digested a little Freud together with a little Bion. He writes: 'Borborygmi may signal the process and acquisition of new thoughts (symbolization) and the free associations derived from borborygmi often provide the key to the understanding of the session by linking the verbal flow of ideas to the underlying sensory and affective experience, thereby providing a "moment of truth". Within the primitive maternal transference, borborygmi are often accompaniments to the fantasy or the hallucination of being fed by the analyst.'

The name Guy DaSilva will be familiar to some readers as the star of hundreds of psychologically gut-wrenching films, among them *Beyond Reality 3*, *The Lube Guy*, *Attack of the Killer Dildos*, and *Porn-O-Matic*

2000. But Guy DaSilva the actor and Guy Da Silva the psychoanalyst are not the same person, no matter how similarly stimulating their work may be.

Müller, Christian (1984). 'New Observations on Body Organ Language.' *Psychotherapy and Psychosomics* 42 (1–4): 124–26.
Da Silva, Guy (1990). 'Borborygmi as Markers of Psychic Work During the Analytic Session. A Contribution to Freud's Experience of Satisfaction and to Bion's Idea About the Digestive Model for the Thinking Apparatus.' *International Journal of Psychoanalysis* 71: 641–59.
— (1998). 'The Emergence of Thinking: Bion as the Link Between Freud and the Neurosciences.' In M. Grignon (ed.) *Psychoanalysis and the Zest for Living: Reflections and Psychoanalytic Writings in Memory of W. C. M. Scott.* Binghamton, N.Y.: ESF Publishers.

THE TASTING OF THE SHREW

If you like shrews, especially if you like them parboiled, you'll want to devour a study published in the *Journal of Archaeological Science*. Called 'Human Digestive Effects on a Micromammalian Skeleton', it explains how and why one of its authors – either Brian D. Crandall or Peter W. Stahl, we are not told which – ate and excreted a ninety-millimetre-long (excluding the tail, which added another twenty-four millimetres) northern short-tailed shrew (species name: *Blarina brevicauda*).

This was, in technical terms, 'a preliminary study of human digestive effects on a small insectivore skeleton', with 'a brief discussion of the results and their archaeological implications'.

Crandall and Stahl are anthropologists at the State University of New York in Binghamton. The shrew was a local specimen, procured via snap trapping at an unspecified location not far from the school.

For the experiment's input, preparation was exacting. After being skinned and eviscerated, the report says, 'the carcass was lightly boiled for approximately 2 minutes and swallowed without mastication in hind and forelimb, head, and body and tail portions'.

Here's how Crandall and Stahl handled the output: 'Faecal matter was collected for the following 3 days. Each faeces was stirred in a pan of warm water until completely disintegrated. This solution was then decanted through a quadruple-layered cheesecloth mesh. Sieved contents were rinsed with a dilute detergent solution and examined with a hand lens for bone remains.' They then examined the most in-

teresting bits with a scanning electron microscope, at magnifications
ranging from 10 to 1000 times.

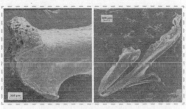

**Digestive damage to a (left) surviving shrew
humerus and (right) surviving shrew tibio-
fibula**

A shrew has lots of bony parts. All of them entered Crandall's
gullet, or maybe Stahl's. But despite extraordinary efforts to find and
account for each bone at journey's end, many went missing. One of the
major jawbones disappeared. So did four of the twelve molar teeth,
several of the major leg and foot bones, nearly all of the toe bones, and
all but one of the thirty-one vertebrae. And the skull, reputedly a very
hard chunk of bone, emerged with what the report calls 'significant
damage'.

The vanishing startled the scientists. Remember, they emphasize
in their paper that this meal was simply gulped down: 'The shrew was
ingested without chewing; any damage occurred as the remains were
processed internally. Mastication undoubtedly damages bone, but the
effects of this process are perhaps repeated in the acidic, churning
environment of the stomach.'

Chewing, they almost scream at their colleagues, is only part of
the story. In each little heap of remains from ancient meals, there be
mystery aplenty.

Prior to this experiment, archaeologists had to, and did, make all
kinds of assumptions about the animal bones they dug up – especially
as to what those partial skeletons might indicate about the people
who presumably consumed them. Crandall and Stahl, through their
disciplined lack of mastication, have given their colleagues something
toothsome to think about.

Crandall, Brian D., and Peter W. Stahl (1995). 'Human Digestive Effects on a Micromammalian
Skeleton.' *Journal of Archaeological Science* 22 (6): 789–97.

MAY WE RECOMMEND

'PHARYNGEAL IRRITATION AFTER EATING COOKED TARANTULA'

by Stephen J. Traub, Robert S. Hoffman, and Lewis S. Nelson
(published in the *Internet Journal of Medical Toxicology*, 2001)

THE WATER TEST

Similar Preference for Natural Mineral Water between
Female College Students and Rats

Yukiko Esumi and Ikuo Ohara*

Shimane Women's Junior College, Matsue 690-0044, Japan
* *Faculty of Home Economics, Kobe Women's University, Suma-ku, Kobe 654-8585, Japan*

The study 'Similar Preference for Natural Mineral Water between
Female College Students and Rats' pulls off a nice bit of interspecies
diplomacy. Reading it end to end, you would be hard pressed to say
who – the women or the rats – was most intended to benefit from the
research.

Written by Esumi Yukiko of Shimane Women's Junior College in
Matsui, Japan, and Ohara Ikuo of Kobe Women's University, and published in the *Journal of Home Economics of Japan*, this six-page monograph
describes an apparently straightforward experiment.

The authors explain that their work was partly inspired by a simple
fact: 'The Society for the Study of Tasty Water, which is sponsored by
the Ministry of Public Welfare, proposed hardness to be one of the
most important requirements for tasty water.' Therefore, they say, 'The
objectives of this study are to investigate the best mineral water for
drinking by using hardness as an index, and whether the response of
rats to mineral water can be extrapolated to that of humans.'

Yukiko and Ikuo conducted taste tests with sixteen healthy female
humans, sixteen healthy female rats, and fourteen different brands
(nine Japanese, two Belgian, and two French) of bottled water. The
water, all of it, was uncarbonated. For good measure, the taste testers
also taste-tested tap water.

The women drank from cups, the rats from objects called 'drinking tubes'. The report specifies that the rats each weighed 160 grams,

give or take three grams, and were 'housed individually'. We are told nothing, not a blessed iota of fact, about the weight of the women, or about their living arrangements. The report also specifies that 'Before beginning the experiment, each animal was fed on a commercial stock diet', but says nothing about what the women consumed.

Yukiko and Ikuo reached two main conclusions. First, they write, 'appropriate levels of minerals are needed for tasty drinking water, too little being as bad as too much, with around 58.3 milligrams/liter of hardness being most favorable'. Second, and perhaps more memorably: 'The present study has demonstrated that the preference for different types of natural mineral water by female college students was similar to that by rats.'

Yukiko and Ikuo make no claim that theirs is the final word. For one thing, they point out, 'The menstrual cycle of the subjects was not considered in this experiment, although taste sensitivity can be influenced by it.'

'Similar Preference for Natural Mineral Water between Female College Students and Rats' is not the only research study to proudly, explicity comparison-test college students and rats. But it may be the most exuberant since C. Lathan and P. E. Fields's 1936 tell-all 'A Report on the Test-retest Performances of 38 College Students and 27 White Rats on the Identical 25 Choice Elevated Maze'.

Yukiko, Esumi, and Ohara Ikuo (1999). 'Similar Preference for Natural Mineral Water between Female College Students and Rats.' *Journal of Home Economics of Japan* 50 (12): 1217 22.
Lathan, C., and P. E. Fields (1936). 'A Report on the Test-retest Performances of 38 College Students and 27 White Rats on the Identical 25 Choice Elevated Maze.' *Journal of Genetic Psychology* 49: 283–96.

EGGS ALLSORTS, BIRDFLESH EVERYKIND

Which birds are the most edible, and which are the least? During and just after World War II, Hugh B. Cott of the University of Cambridge doggedly pursued these questions, using means that were waspy, feline, and human. His discoveries are summed up in a 154-page report entitled 'The Edibility of Birds – Illustrated by 5 Years Experiments and Observations (1941–1946) on the Food Preferences of the Hornet, Cat and Man'.

In October 1941, Cott made a chance observation. While collecting and preserving bird skins in Beni Suef, Egypt, he discarded the meaty parts of a palm dove (*Streptopelia senegalensis aegyptiaca*) and a pied kingfisher (*Ceryle rudis rudis*). Hornets descended upon the palm dove carcass, but ignored the kingfisher.

Cott, entranced, later offered other hornets a choice of different cuts (breast, wings, legs, and gut) of some forty different bird meats, in 141 experiments conducted in Beni Suef, Cairo, and Tripoli, Lebanon.

The hornets especially took to crested lark, greenfinch, white-vented bulbul, and house sparrow. They voted (metaphorically) thumbs down on golden oriole, hooded chat, masked shrike, and hoopoe, among others.

Cott conducted another forty-eight experiments, with nineteen kinds of bird meat, using three cats (two in Cairo, one in Tripoli) as tasters. In each experiment, the taster chose (or chose not to choose) between two different bird meats.

To answer the 'which would a human eat' question, Cott gathered data 'from natives in the Lebanon; from personal experience and from observations sent in reply to a published inquiry; and from the [scientific] literature'. He drew most heavily from Reverend H. A. Macpherson's occasionally mouthwatering 1897 tome *A History of Fowling*.

Surveying the results of all those taste tests of all those birds by hornets, cats, and people, Cott saw both rhyme and reason. He concluded that, in most cases, humans and cats 'agreed with the hornets in rating more conspicuous species as relatively distasteful when compared with more cryptic species ... Birds which are relatively vulnerable and conspicuous ... appear in general to be more or less highly distasteful – to a degree likely to serve as a deterrent to most predators.'

At the other extreme, birds that have especially inconspicuous or camouflaged appearance, Cott almost cackles, 'are also those which are especially prized for the excellence of their flesh'. The list of these includes the Eurasian woodcock, skylark, and the mallard duck.

Among the widely disliked were kingfishers, puffins, and bullfinches. Cott cautioned his readers that 'palatability may change with growth and age of the bird; and it differs markedly in different parts of the same individual'.

But as with the special case of chickens and eggs, this is neither the beginning of the story nor its end. At roughly the same time, Cott was also running an extensive programme to test the palatability of every kind of bird egg he could find. The titles of his studies are pretty self-explanatory:

> The Palatability of the Eggs of Birds – Illustrated by Experiments on the Food Preferences of the Hedgehog (*Erinaceus europaeus*)
>
> The Palatability of the Eggs of Birds: Illustrated by Three Seasons' Experiments (1947, 1948 and 1950) on the Food Preferences of the Rat (*Rattus norvegicus*)
>
> The Palatability of the Eggs of Birds – Illustrated by Experiments on the Food Preferences of the Ferret (*Putorius furo*) and Cat (*Felis catus*) – with Notes on Other Egg-Eating Carnivora' (Those other carnivora are numerous, and include civets, mongooses and meerkats, hyenas, dogs and dingoes, otters, aardwolves, and foxes.)

Cott's research programme could (although was not, so far as I know) be summarized as 'Go suck eggs!'

Egg palatability experiments are potentially of great practical value. Island nations, Britain pre-eminently, were and are vulnerable to enemies who would block food shipments from overseas. One could counter the danger by discovering unknown or unappreciated edible native foodstuffs. A simple way to begin: collect bird eggs and test their palatability.

Egg collecting, like other research activities, is not without hazards. Cott relates an incident that's documented in an 1882 monograph: 'The victim, having collected a basket-full of the first eggs of the season, and wishing to procure more, had sent his wife to empty the basket in the village. In her absence, he fixed his rope to the cliff-top and made a second descent. Meanwhile a fox ran up, and gnawed the rope till it severed, at the place where the man had previously rubbed his yolk-smeared hands.'

Cott's experiments mainly addressed a scientific question – demonstrating that, usually, the most conspicuous eggs taste terrible to whatever might want to eat them.

Cott also used human egg taste-testers. In 1946, he entered a six-year collaboration with the Cambridge Egg Panel, one of many similar bodies formed during World War II to help regulate the nation's food supply. Under Cott's direction, panellists tasted the eggs of 212 bird species. This resulted in a 129-page report called 'The Palatability of Eggs and Birds: Mainly Based upon Observations of an Egg Panel'. It has raw data, supplemented with colourful highlights from the tasters' own notes and from other sources, including Cott's coterie of egg-collecting correspondents.

For the egg panellists, 'samples were tested in the form of a scramble, prepared over a steam-bath, without any addition of fat or condiment'. Each taster assessed each sample on a scale dropping from 'ideal' way down to 'repulsive and inedible'.

The paper concludes with a list of the different egg types 'in descending order of acceptability'. Keep in mind that these are the aggregate preferences; individual tastes may vary. Most acceptable: chicken, then emu and coot, then black-backed gull. The eggs of last resort, as rated by official British egg-tasting persons: green woodpecker, Verreaux's eagle owl, wren, speckled mousebird, and, dead last, black tit.

Cott's work proved to be, among other things, inspirational. A generation later, Richard Wassersug (see page 170) cited it as both inspiration and, to some extent, guide for his research into the palatability of Costa Rican tadpoles.

Cott, Hugh B. (1945). 'The Edibility of Birds.' *Nature* 156 (3973): 736–37.

— (1947). 'The Edibility of Birds – Illustrated by 5 Years Experiments and Observations (1941–1946) on the Food Preferences of the Hornet, Cat and Man – and Considered with Special Reference to the Theories of Adaptive Coloration.' *Proceedings of the Zoological Society of London* 116 (3–4): 371–524.

— (1948). 'Edibility of the Eggs of Birds.' *Nature* 161 (4079): 8–11.

— (1951). 'The Palatability of the Eggs of Birds - Illustrated by Experiments on the Food Preferences of the Hedgehog (*erinaceus-europaeus*).' *Proceedings of the Zoological Society of London* 121 (1): 1–40.

— (1952). 'The Palatability of the Eggs of Birds: Illustrated by Three Seasons' Experiments (1947, 1948 and 1950) on the Food Preferences of the Rat (*Rattus norvegicus*); and with Special Reference to the Protective Adaptations of Eggs Considered in Relation to Vulnerability.' *Proceedings of the Zoological Society of London* 122 (1): 1–54.

— (1953). 'The Palatability of the Eggs of Birds – Illustrated by Experiments on the Food Preferences of the Ferret (*Putorius-Furo*) and Cat (*Felis-Catus*) – With Notes on Other Egg-Eating Carnivora.' *Proceedings of the Zoological Society of London* 123 (1): 123–41.

— (1954). 'The Palatability of Eggs and Birds: Mainly Based upon Observations of an Egg Panel.' *Proceedings of the Zoological Society of London* 124 (2): 335–463.

A BEEF BOOM

Before John Long applied his expertise to the problem, people tried many ways to make meat more tender – chewing it, pounding it, soaking it in enzymes.

The report, 'Hydrodyne Exploding Meat Tenderness', published in 1998 by the US Department of Agriculture (USDA), describes Long's act of creation as 'a peacetime use for explosives'. It goes on: 'Throughout John Long's career as a mechanical engineer, he worked with explosives at Lawrence Livermore [National Laboratory]. His mission: preparing the Nation's defense. He always wondered if the explosives he studied could be used for peaceful ends – like tenderizing meat. Then, after more than 10 years of retirement and long after the Cold War's end, he began pursuing the Hydrodyne concept in earnest.'

The article explains that in 1992, Long teamed up with a meat scientist, Morse Solomon. Their first set-up was 'an ordinary plastic drum filled with water and fitted with a steel plate at the bottom to reflect shock waves from an explosion'. By 1998, Long and Solomon were stuffing meat, water, and explosives into a seven-thousand-pound (3180-kilogram) steel tank covered with an eight-foot (2.4-metre) steel dome.

This official USDA story of how it all began looks past the fact that another man, Charles Godfrey of Berkeley, California, obtained a patent in 1970 for his 'apparatus for tenderizing food'. Godfrey's first sentence blasts away all confusion: 'An article of food is tenderized by placing it in water and detonating an explosive charge in the vicinity thereof.'

Godfrey explains his method: 'A cut of meat desired to be tenderized is placed under water within a tank. In view of the tendency of the meat to float, it may be necessary to tie the meat in position by a string ... A compressive pressure wave traveling at a speed higher than the velocity of sound may be generated in the water by a means, such as a charge of high explosive, which is supported above the meat by any suitable means, such as the leads which are used to ignite the detonator of the high explosive.'

Once the idea was out there, other scientists took to experimenting with beef, pork, chicken, and other things that went boom. A study in 2006 alluded to a scientist named Schilling who showed that 'the hydrodynamic shock wave ... did not affect the color of cooked broiler breast meat'.

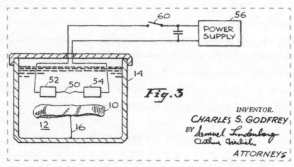

How to generate a shock pressure wave for tenderizing an article of food. Detail from US Patent no. 3,492,688

A pamphlet from the National Cattlemen's Beef Association bragged that 'technology has been shown to improve the tenderness of beef by 30-80% and, the tougher the piece of meat, the greater the magnitude of improvement'. But so far the process works well only for small, sub-industrial quantities. The niggling problem, when applying explosives to heaps of flesh, is how to tenderize without pulverizing.

Lee, Jill (1998). 'Hydrodyne Exploding Meat Tenderness.' *Agricultural Research* (June): 8–10.
Godfrey, Charles S. (1970). 'Apparatus for Tenderizing Food.' US Patent no. 3,492,688, 3 February.

PET PALATES

Pet food taste-testing by humans rose to a new level of formality with the publication of the scholarly study 'Optimizing the Sensory Characteristics and Acceptance of Canned Cat Food: Use of a Human Taste Panel'. The author, G. J. Pickering of Brock University in St Catharines, Ontario, Canada, reports: 'Cats are sensitive to flavour differences in diet, very discriminative in food selection, and clearly unable to verbalize their likes and dislikes. These issues have dogged the industry for decades.'

Pickering explains that taste tests with volunteer cats suffer three drawbacks. They are 'expensive to maintain, time consuming, and yield limited and often equivocal data'. So he offers an alternative: 'In-house tasting trials using a human taster are commonly conducted by the pet food industry, although there is a paucity of relevant information in the scientific literature.'

His study serves up a hearty helping of information. Human volunteers rated thirteen different commercial pet food samples, concentrating on eighteen so-called flavour attributes: sweet, sour/acid, tuna, herbal, spicy, soy, salty, cereal, caramel, chicken, methionine, vegetable, offaly, meaty, burnt, prawn, rancid, and bitter.

Product Descriptions

Product code	Product description
A	Homogeneous product marketed as minced-beef based
B	Homogeneous product marketed as jelly-meat based
C1	Product C – binary system: meat chunks portion
C2	Product C – binary system: gravy portion
D	Homogeneous product
E	Homogeneous product
F1	Product F – binary system: meat chunks portion
F2	Product F – binary system: gravy portion
G	Homogeneous product marketed as fish based
H1	Product H – binary system: meat chunks portion
H2	Product H – binary system: gel portion
I1	Product I – binary system: meat chunks portion
I2	Product I – binary system: gravy portion

The tasting protocols depended on the texture of what was being tasted. When munching on meat chunks people assessed the hardness, chewiness, and grittiness ('sample chewed using molars until masticated to the point of being ready to swallow'). But they gauged gravy/gel glops for viscosity and grittiness ('sample placed in mouth and moved across tongue').

The knowledge thus gained is only a first step. 'It is now necessary', Pickering writes, 'to determine the usefulness and limits of sensory data gathered from human panels in describing and predicting food acceptance and preference behaviours in cats.'

Where the Pickering cat food paper was mainly for industrial consumption, a team of independent scholars – comprised of John Bohannon, Robin Goldstein, and Alexis Herschkowitsch – published 'Can People Distinguish Pâté From Dog Food?' to address a societal concern: 'the potential of canned dog food for human consumption by assessing its palatability alone'. The study concludes somewhat perplexedly that (1) 'human beings do not enjoy eating dog food' and (2) are 'not able to distinguish its flavor profile from other meat-based products that are intended for human consumption'.

Perhaps alcohol helped this to happen. The dog monograph is published by the American Association of Wine Economists (AAWE), while the cat paper is written by a professor of biological sciences/wine science, and appears in the *Journal of Animal Physiology and Animal Nutrition*, which serves up complementary studies such as 'The Influence of Polyphenol Rich Apple Pomace or Red-Wine Pomace Diet on the Gut Morphology in Weaning Piglets'.

Pickering, G. J. (2009). 'Optimizing the Sensory Characteristics and Acceptance of Canned Cat Food: Use of a Human Taste Panel.' *Journal of Animal Physiology and Animal Nutrition* 93 (1): 52–60. Bohannon, John, Robin Goldstein, and Alexis Herschkowitsch (2007). 'Can People Distinguish Pâté From Dog Food?' AAWE Working Paper no. 36, April.

MEASURED ATTITUDES TO CHOCOLATE

A report called 'The Development of the Attitudes to Chocolate Questionnaire', published in 1998, tells how three researchers at the University of Wales, Swansea, cooked up a new analytic tool.

Psychologists had long craved a way to assess someone's craving for chocolate. Why chocolate? Because 'chocolate is by far the most commonly craved food'. It tempts chocoholics, and also academics who hunger for knowledge and perhaps recognition.

The desired goal – the perhaps impossible dream – is to measure and compare any two people's chocolate cravings as reliably as one can measure and compare the heights of two tables. But cravings are often intertwined with emotions, and table heights are not. This explains why table heights are easier to measure.

All prior attempts to measure cravings, say study co-authors, David Benton, Karen Greenfield, and Michael Morgan, were 'unreliable'. They devised a tool that, they say, 'provides a quantitative estimate of the fundamental attitudes to chocolate'. It measures the magnitude of the craving; it also measures the guilt feelings.

The tool – called the Attitudes to Chocolate Questionnaire – is a simple list of twenty-four statements. Some are strictly about craving:

> *The thought of chocolate often distracts me from what I am doing.*
> *My desire for chocolate often seems overpowering.*

Some are about guilt:

> *I feel guilty after eating chocolate.*
> *After eating chocolate I often wish I hadn't.*

To measure an individual's chocolate craving, have the person read each statement and then indicate, by marking on a little ruled line, whether the statement is 'not at all like me' or 'very much like me' or somewhere in between. A bit of statistical manipulation, and *hey, presto!* out pops a set of numbers that describe the craving.

Benton, Greenfield, and Morgan tested and calibrated their new tool on some student volunteers. In addition to answering the questionnaire, the volunteers played a sort of mechanical game. By pressing a lever, they could obtain a reward – a little button made of chocolate. As the game progressed, they had to press the bar more and more times (twice, then four times, then eight, then sixteen, etc.) before another chocolate would pop out. The point at which someone refused to keep playing this game indicated the strength of their craving (or, one might argue after some time has passed, the fullness of their stomach).

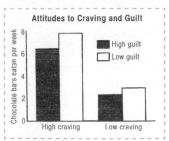

Figure: 'The relationship between craving, guilt, and the eating of chocolate bars ... High craving but not guilt was associated with the eating of a higher number of chocolate bars'

Afterwards, the researchers compared people's scores on the Attitudes to Chocolate Questionnaire with their craving strength as measured in the press-the-lever-and-get-a-treat game.

The questionnaire results accorded well with what happened in the game. In most cases, if the questionnaire said someone had a high craving for chocolate, that person was persistent at making the frustrating machine deliver up chocolates. Thus, the little questionnaire is a cheap, fairly accurate way to measure chocolate craving and also to measure guilt.

In their report, the researchers announce that, using their new tool, they made an exciting new psychological discovery: that 'craving but not guilt was associated with the eating of chocolate bars'.

Benton, David, Karen Greenfield, and Michael Morgan (1998). 'The Development of the Attitudes to Chocolate Questionnaire.' *Personality and Individual Differences* 24 (4): 513–20.

Cramer, Kenneth M., and Mindy Hartleib (2001). 'The Attitudes to Chocolate Questionnaire: A Psychometric Evaluation.' *Personality and Individual Differences* 31 (6): 931–42.

PAH TO WHISKY

Whisky and candlelight, consumed repeatedly over many years, involve some measure of danger. Two Dutch research projects tried to take that measure. They hoped to confront the spectre of death – to either confirm or disprove the worry that good whisky and sacred candles, singly or in combination, are very, very bad for a body.

The very specific object to be measured, both in the booze and in the candle smoke, was a particular group of chemicals. Known by the acronym 'PAHs' (say it aloud, with pursed lips, to see how disgustingly nasty some people think they are) – these tasty, smelly molecules have a fairly well-deserved reputation for causing cancer and other illnesses.

Jos Kleinjans, an environmental health professor at the University of Maastricht, led a pair of inquisitions. He joined with one bunch of colleagues to give whisky a good going-over. With a different bunch, he sniffed into church candle (and also church incense) fumes, in search of insidious nastiness.

The whisky came first. As detailed in a report published in 1996 in the *Lancet*, Kleinjans and five friends obtained some of the finest whiskies on Earth. For comparative purposes, they also picked up some of the cheap stuff.

From Scotland they got six malts – Laphroaig, Oban, Glenkinchie, Glenfiddich, Highland Park, and Glenmorangie – and also four blends – Famous Grouse, Chivas Regal, Johnnie Walker Red, and Ballantines.

From North America, five bourbons – Southern Comfort, Virginia Gentleman, Jack Daniel's, Four Roses, and Old Overholt.

From Ireland, three whiskies – Bushmill's Malt, Jameson, and Tullamore Dew.

'Carcinogenic PAHs', the scientists announced, 'were present in all whisky brands' but 'it is apparent that Scotch malts have the highest carcinogenic potential'. Eye-pokingly, they revealed that the most expensive Scotch malts contain the highest levels of danger.

No worries, though, or at least not many. The report concludes: 'Compared with smoked and char-broiled food products ... PAH concentrations in whiskies are low, and are not likely to explain the cancer risks of whisky consumption.' This is danger with a most tiny 'd'. It is the spice of life, and also of whisky.

An almost biblical seven long years later, Kleinjans and three other friends published a report called 'Radicals in the Church'. It tells of their adventures in a Roman Catholic Church – the Onze Lieve Vrouwe Basiliek in Maastricht. There, they sampled the fumes from a standard (to the extent that such things are standard) nine-hour session of burning candles and incense. They also, literally for good measure, sampled the air before and after what they call a 'simulated service' in a large basilica. The PAH levels, they discovered, are higher than in a dose of whisky, but perhaps not high enough to shed clear light on the question 'is it dangerous?'

And so their report ends with a murky diagnosis: 'It cannot be excluded that regular exposure to candle- or incense-derived particulate matter results in increased risk of lung cancer or other pulmonary diseases.'

Kleinjans, Jos C, S., Edwin J. C. Moonen, Jan W. Dallinga, Harma J. Albering, Anton E. J. M. van den Bogaard, and Frederik-Jan van Schooten (1996). 'Polycyclic Aromatic Hydrocarbons in Whiskies.' *Lancet* 348: 1731.

De Kok, T. M. C. M., J. G. F. Hogervorst, J. C. S. Kleinjans, and J. J. Briede (2004). 'Radicals in the Church.' *European Respiratory Journal* 24: 1–2.

STANDARD FOOD GLOPS

When food manufacturers put nutrition info on their labels, they can either (a) invent the numbers (and risk going to prison) or (b) chemically analyse the food to see how much of it is saturated fat, or sodium, or vitamin A, or some other particular nutrient, mineral, or vitamin. The analytical chemists, if they are honest and honourable, must know whether they can trust their own measurements – and so they test their equipment by first analysing some officially measured and certified 'typical' foodstuff.

For just $839 (£534) one can buy the essence of an officially measured and certified 'typical diet' – officially prepared and bottled by the US government's National Institute of Standards and Technology (NIST). The money gets you twelve grams of blended, 'freeze-dried homogenate of mixed diet foods', delivered in a pair of six-ounce bottles.

An accompanying NIST document, called 'Certificate of Analysis, Standard Reference Material 1548a, Typical Diet', makes no claims as to tastiness. The certificate notes that these possibly delicious dollops are 'not for human consumption'.

Each portion contains a soupçon of mystery, a hint of inexactitude in its numbers. The Certificate of Analysis makes mention of 'uncertainties that may reflect only measurement precision, may not include all sources of uncertainty, or may reflect a lack of sufficient statistical agreement among multiple methods'. (The certificate goes on to mention, with a metaphorical twirling of its moustache and twinkling of its eyes, that 'there is insufficient information to make an assessment of the uncertainties'.)

Despite the imprecision, it would be wrong, very wrong, to say that the diet is slopped together carelessly. To the contrary, it was 'prepared from menus used for the metabolic studies at the Human Study Facility' of the US Food and Drug Administration. 'Food items in prescribed quantities representing a four-day menu cycle were pooled/combined into a master menu ... The material was freeze-dried, pulverized, sieved, and radiation sterilized at a dose of 2.5 mrad to prevent bacterial growth', then 'blended, bottled, and sealed under nitrogen'.

In addition to the typical diet, NIST produces items conceivably of appeal to more specialized palates: baby food composite, peanut butter, baking chocolate, meat homogenate, and Lake Superior fish tissue. The latter includes standard amounts of fat, fatty acid, pesticides, polychlorinated biphenyls (PCBs), mercury, and methylmercury.

NIST offers many kinds of useful and, to the connoisseur, delightful Standard Reference Materials. Their catalogue runs to 145 pages.

Prospective purchasers can peruse page after page of bodily fluids and glops, among them bilirubin, cholesterol, and ascorbic acid in frozen human serum. There are other specialty products in dizzying variety: toxic metals in bovine blood, naval brass, domestic sludge, and plutonium-242 solution, to name four.

Prices are mostly in the $300 to $600 range. At the high end, you will find New York/New Jersey waterway sediment for $610. There are bargains to be had, including an item called 'cigarette ignition strength, standard', on offer at one carton (two hundred cigarettes) for $192. Alas, 'multi drugs of abuse in urine' was out of stock, the last time I looked.

National Institute of Standards and Technology (2009). 'Certificate of Analysis – Standard Reference Material 1548A: Typical Diet,' https://www-s.nist.gov/srmors/view_detail.cfm?srm=1548A. Sharpless, Katherine E., Jennifer C. Colbert, Robert R. Greenberg, Michele M. Schantz, and Michael J. Welch (2001). 'Recent Developments in Food-Matrix Reference Materials at NIST.' *Fresenius Journal of Analytic Chemistry* 370: 275–78.

IN BRIEF

'DISTINCTION BETWEEN HEATING RATE AND TOTAL HEAT ABSORPTION IN THE MICROWAVE-EXPOSED MOUSE'

by Christopher J. Gordon and Elizabeth C. White (published in *Physiological Zoology*, 1982)

According to the authors at the US Environmental Protection Agency, 'This investigation assesses the ability of the heat-dissipating system of the mouse to respond to equivalent heat loads (e.g., J/g) administered at varying intensities (e.g., J/g/s or W/kg). Use of a microwave exposure system provided a means to administer exact amounts of energy at varying rates in awake, free-moving mice.'

IMPLICATIONS OF CUSTARD

There is one individual who, above all others, has plumbed the effects of custard.

René A. de Wijk is based at the Wageningen Centre for Food Sciences in the Netherlands. During a four-year burst of scholarship, demonstrating his stunning productivity, de Wijk published more than ten custard-centric research reports, each of them a substantial contribution to our understanding of, and relationship with, custard.

Do not think of custard researchers as being solitary, asocial creatures. Certainly de Wijk is not. He shares co-authorship credit with a happy variety of colleagues.

In 2001, de Wijk teamed up with H. Weenen, L. J. Van Gemert, R. J. M. Van Doorn, and G. B. Dijksterhuis. The result was 'Texture and Mouthfeel of Semi-Solid Foods: Commercial Mayonnaises, Dressings, Custard Desserts and Warm Sauces', which delighted readers of the *Journal of Texture Studies*.

Two years later, de Wijk, together with Weenen and two others, published a pudding-studies instant classic, 'The Influence of Bite Size and Multiple Bites on Oral Texture Sensations'. This carefully worded document describes a pair of experiments.

First, the scientists observed what happens when a person takes carefully measured bites of a vanilla custard dessert. Eating custard in single bites, they observed, 'affected perception of thickness, temperature, astringency, and creaminess'. In the other experiment, the custard-chewing volunteers began taking bites from one vanilla custard dessert – but then suddenly switched to biting an entirely different vanilla custard dessert. The effect was fairly subtle: 'sensations of thickness and fatty afterfeel' became more noticeable.

In 2003, de Wijk and colleagues issued two reports about the interaction of saliva and custard. In one, they tested the effect of adding saliva to custard prior to eating that custard. The report carefully notes that 'saliva had previously been collected from the subjects and each subject received his/her own saliva'. The other report looked at 'whether and how the amount of saliva a subject produces influences the sensory ratings' when that person then gobbles a vanilla custard dessert. The results are summarized memorably: 'A subject with a larger saliva flow rate during eating did not rate the foods differently from a subject with less saliva flow.'

Another de Wijk report from that year explored the effects of manipulating custard inside one's mouth. The activities 'ranged from simply placing the stimulus on the tip of the tongue to vigorously moving it around in the mouth'. To gain some perspective, the test subjects also had to manipulate mayonnaise, although that was done separately.

Amount of ingested custard dessert as affected
by its color, odor, and texture

René A. de Wijk[a,b,*], Ilse A. Polet[a,c], Lina Engelen[a,d],
Rudi M. van Doorn[a,d], Jon F. Prinz[a,d]

Abstract

The effects of nonoral sensations, such as visual texture and odor, on the size of the first bite were investigated in a series of studies using specially constructed food delivery cups with lower, firm which custards were ingested ("ingested custard"), and upper, from which a custard was viewed and/or smelled ("upper custard") compartments. Ingested and upper custards were either the same or different. Bite size

De Wijk, together with four colleagues, then came out with his magnum opus, a distillation of what is known about the sensation of mouthing custard and an intellectually, gustatorially stimulating read. For some readers, 'Amount of Ingested Custard Dessert as Affected by Its Color, Odor, and Texture' will recall the work of Marcel Proust, for it deals entirely with what happens when a sensitive human being takes the very first bite of custard.

Weenen, H., L. J. Van Gemert, R. J. M. Van Doorn, G. B. Dijksterhuis, and R. A. de Wijk (2001). 'Texture and Mouthfeel of Semi-Solid Foods. Commercial Mayonnaises, Dressings, Custard Desserts and Warm Sauces.' *Journal of Texture Studies* 34 (2), 159–79.

De Wijk, R. A., L. Engelen, J. F. Prinz, and H. Weenen (2003). 'The Influence of Bite Size and Multiple Bites on Oral Texture Sensations.' *Journal of Sensory Studies* 18 (5): 423–35.

Engelen, L., R. A. de Wijk, J. F. Prinz, A. M. Janssen, H. Weenen, and F. Bosman (2003). 'A Comparison of the Effects of Added Saliva, Alpha-Amylase and Water on Texture Perception in Semi-Solids.' *Physiology and Behavior* 78: 805–11.

Engelen, L., R. A. de Wijk, J. F. Prinz, A. Van der Bilt, and F. Bosman (2003). 'The Relation between Saliva Flow after Different Stimulations and the Perception of Flavor and Texture Attributes in Custard Desserts.' *Physiology and Behavior* 78 (1): 165–69.

De Wijk, R. A., L. Engelen, and J. F. Prinz (2003). 'The Role of Intra-Oral Manipulation on the Perception of Sensory Attributes.' *Appetite* 40 (1): 1–7.

De Wijk, R. A., et al. (2004). 'Amount of Ingested Custard Dessert as Affected by its Color, Odor, and Texture.' *Physiology and Behavior* 82 (2–3): 397–403.

Janssen, A. M., Marjolein E. J. Terpstra, R. A. de Wijk, and J. F. Prinz (2007). 'Relations Between Rheological Properties, Saliva-induced Structure Breakdown and Sensory Texture Attributes of Custards.' *Journal of Texture Studies* 38 (1): 42–69.

THE GARLICKY FAMILY

'This study assessed the effects of the odor and ingestion of garlic bread on family interactions.' With those words, Alan R. Hirsch of the Smell & Taste Treatment and Research Foundation, in Chicago, declared the purpose and the breadth of his research. However, Hirsch did not analyse the matter as deeply as he could have.

EFFECTS OF GARLIC BREAD ON FAMILY INTERACTIONS
Alan R. Hirsch, Smell & Taste Treatment and Research Foundation,
Chicago, IL

This study assessed the effects of the odor and ingestion of garlic bread on
family interactions.
Fifty families were given 2 identical spaghetti dinners, randomly presented
with and without garlic bread. Average family size was 3.6 (range 2 to 12).
At each dinner, the number of interactions both positive and negative were
recorded during 3 one-minute intervals. The 1st minute served as a
baseline. During the 2nd minute the garlic bread aroma was presented.
During the 3rd minute the bread was ingested. Three minutes of interactions

This is not to say that Dr Hirsch was lazy. His experiment examined the interactions of garlic bread and fifty families, an undertaking that involved the preparation and consumption of not just fifty, but a full one hundred meals. Each family was made to experience dinner with garlic bread, and also dinner without. For each family, the order of those two experiences was determined randomly.

Hirsch published details in the journal *Psychosomatic Medicine*. The families ranged in size from two to twelve people. In their breaded meal, each family had to endure a full minute before being exposed to the garlicky aroma. Hirsch's published account reads like the science adventure tale it is. 'During the second minute', he writes, 'the garlic bread aroma was presented. During the [third] minute the bread was ingested.'

The rest of the story can and is told in numbers. 'Smelling and eating garlic bread decreased the number of negative interactions between family members', the report says, and 'the number of pleasant interactions increased.' Hirsch reached the conclusion that: 'Serving garlic bread at dinner enhanced the quality of family interactions. This has potential applications in promoting and maintaining shared family experiences, thus stabilizing the family unit, and also may have utility as an adjunct to family therapy.'

But what, biochemically, is the mechanism for this effect? On that level, Hirsch is mum.

For an answer, one must look elsewhere, perhaps to the *Journal of Biological Chemistry*, which published a study called 'The Active Principle of Garlic at Atomic Resolution'. The German authors of that report caution that 'despite the fact that many cultures around the world value and utilize garlic as a fundamental component of their cuisine as well as of their medicine cabinets, relatively little is known about the plant's protein configuration that is responsible for the specific properties of garlic.'

This scarcity of knowledge also obtruded itself in 1998, when three scientists in Wales published 'What Sort of Men Take Garlic Preparations?' Their conclusion: 'Men who take garlic supplements are generally similar to non-garlic users.'

Hirsch, Alan R. (2000). 'Effects of Garlic Bread on Family Interactions.' *Psychosomatic Medicine* 62 (1): 103.

Kuettner, E. Bartholomeus, Rolf Hilgenfeld, and Manfred S. Weiss (2002). 'The Active Principle of Garlic at Atomic Resolution.' *Journal of Biological Chemistry* 277 (48): 46402–7.

Thomas, H. F., P. M. Sweetnam, and B. Janchawee (1998). 'What Sort of Men Take Garlic Preparations?' *Complementary Therapies in Medicine* 6: 195–97.

TEABAGGING IN THE NAME OF SCIENCE

Political teabagging and sexual teabagging have attracted lots of controversial attention in recent years, but a lesser-known variety – research teabagging – has much to recommend it.

In case you have not encountered the word 'teabagging', here's some linguistic background. Political teabagging takes its name from a twisted, angry dip into American/British history: the 'Boston Tea Party' anti-tax protest of 1773, while sexual teabagging involves dipping one particular body part into another, a bit like a teabag is dipped in a mug.

Research teabagging, in contrast, confronts rather different matters – using teabags to explore scientific and medical questions.

In 2009, a group of nine Japanese researchers told how they used bags of green tea to fight a disgusting odour that arises from the hands of extremely unlucky stroke victims. Their report, published in the journal *Geriatrica and Gerontology International*, 'Four-Finger Grip Bag with Tea to Prevent Smell of Contractured Hands and Axilla in Bedridden Patients', found that clutching a bag filled with green tea 'could substantially control smell in these bedridden patients'.

In 1997, a nurse clinician in Winnipeg, Canada, published a report in the *Journal of Obstetric, Gynecologic and Neonatal Nursing* called 'Does Application of Tea Bags to Sore Nipples While Breastfeeding Provide Effective Relief?' This is a happy story, concluding that 'warm water or tea bag compresses are an inexpensive, equally effective treatment' that 'can prevent further complications such as severe pain, cracking, bleeding, inadequate milk ejection, and, ultimately, premature weaning.'

Seven years later, an American medical team reported a case of drug abuse via teabags. Their paper, called 'The Fentanyl Tea Bag', appeared in the journal *Veterinary and Human Toxicology*. It describes 'a 21-year-old woman who steeped a fentanyl patch in a cup of hot water and then drank the mixture. Coma and hypoventilation resulted'.

Another group of teabaggers used maggots. A 2009 issue of *Turkiye Parazitoloji Dergisi* (the *Turkish Parasitology Digest*) featured a monograph entitled 'The Treatment of Suppurative Chronic Wounds with Maggot Debridement Therapy'. It tells how 'sterile maggots, produced in university laboratories and by private industry, are usually applied to the wound either by using a cage-like dressing or a tea bag-like cage'.

More than thirty years before that, a team of biomedical teabaggers took aim at brown dog ticks. Their 1974 study in the *Bulletin of Epizootic Diseases of Africa* assessed the 'teabag method', using a teabag-like structure filled with maggots for 'testing acaricide susceptibility of the brown dog tick *rhipicephalus sanguineus*'.

Unlike the other forms of teabagging, which involve an element of exhibitionism, research teabagging is a quiet endeavour, typically conducted in low-key fashion, in laboratories or hospitals.

Of the bunch, it's the only one that's typically accompanied and lubricated by many, many cups of actual, teabag-brewed tea.

Kigaye, M. K., and J. G. Matthysse (1974). 'Testing Acaricide Susceptibility of the Brown Dog Tick *Rhipicephalus sanguineus* (Latreille, 1806). II Teabag Method.' *Bulletin of Epizootic Diseases of Africa* 22 (3): 279–85.

Mumcuoglu, Kosta Y., and Aysegul Taylan Ozkan (2009). 'The Treatment of Suppurative Chronic Wounds with Maggot Debridement Therapy.' *Turkiye Parazitoloji Dergisi* 33 (4): 307–15.

Fukuoka, Yumiko, Hisashi Kudo, Aiko Hatakeyama, Naomi Takahashi, Kayoko Satoh, Naoko Ohsawa, Mayumi Mutoh, Masahiko Fujii, and Hidetada Sasaki (2009). 'Four-Finger Grip Bag with Tea to Prevent Smell of Contractured Hands and Axilla in Bedridden Patients.' *Geriatrica and Gerontology International* 9 (1): 97–99.

Fermin Barrueto, Mary Ann Howland, Robert S. Hoffman, and Lewis S. Nelson (2004). 'The Fentanyl Tea Bag.' *Veterinary and Human Toxicology* 46 (1): 30–31.

Lavergne, Noelie A. (1997). 'Does Application of Tea Bags to Sore Nipples While Breastfeeding Provide Effective Relief?' *Journal of Obstetric, Gynecologic and Neonatal Nursing* 26 (1): 53–58.

Brennan, Mike, Janet Hoek, and Philip Gendall (1998). 'The Tea Bag Experiment: More Evidence on Incentives in Mail Surveys.' *International Journal of Market Research* 40 (4): 347–52.

IN BRIEF

'IMPACTION OF AN INGESTED TABLE FORK IN A PATIENT WITH A SURGICALLY RESTRICTED STOMACH'

by A. Cassaro and M. Daliana (published in the *New York State Journal of Medicine*, 1992)

WASCALLY WABBIT WRAPPING

There are few peer-reviewed papers on the subject of designing and testing an improved packaging for large hollow chocolate bunnies. Of these articles, the most bouncily thorough is one called 'Designing and Testing an Improved Packaging for Large Hollow Chocolate Bunnies'. Though just seven pages long, it contains everything a research report ought to have.

The opening section describes the nature of the problem: 'To test the properties required for the packaging of hollow chocolate Easter bunnies to resist any hazards in the distribution environment.' The concluding section suggests that more research is needed.

The experiments are described in clear, spare prose, as are the materials ('The product for our tests was a hollow milk chocolate figure with the shape of an Easter Bunny'), the testing equipment ('The drop testing machine had two drop leaves controlled with a foot paddle'), and the procedures ('Each series of nine bunnies per design was divided into three sets each of three packed bunnies'). At the end comes a list of references, one of which is C. M. Harris's gently moving benchmark 'Shock and Vibration Handbook'.

Drop Heights		
Weight, pounds	Drop height, inches	Type of handling
1-20	30	One-man throw
21-40	24	One-man carry
41-60	18	Two-men carry
61-100	12	Light equipment
Over 100	Incline test	Heavy equipment

The paper is visually informative, with four charts and seven technical renderings. The eye is drawn to Figure 7, a perspective drawing of a chocolate bunny. The bunny is wearing an apron and holding a carrot, and has no legs. The ears point straight up. The facial expression is enigmatically bland, suggesting both Mona Lisa and a mid-career clerk, while resembling neither.

The bunny-packaging scientists, G. M. Greenway and R. E. Garcia Via of the University of Missouri-Rolla's package sealing laboratory, list their results and discuss their conclusions. Commendably, they identify the study's limitations, especially the main one, that 'availability of materials – especially bunnies – was a constraint during this experiment'.

Although there are few peer-reviewed papers on the subject of designing and testing an improved packaging for large hollow chocolate bunnies, there is a considerable body of published research concerning other problems in the discipline of packaging. Want a good introduction to the chemical physics of plastic bags? P. M. Vilela and a colleague at Pontificia Universidad Católica del Perú, in Lima, published a corker in the *European Journal of Physics* in 1999. It involves deformation, spaghetti, Boltzmann's superposition principle, non-linear least-squares fits to the viscous creep, gentle wriggles, and a warmly satisfying title: 'Viscoelasticity: Why Plastic Bags Give Way When You Are Halfway Home'.

Greenway, G. W., and R. E. Garcia Via (1977). 'Designing and Testing an Improved Packaging for Large Hollow Chocolate Bunnies.' *TAPPI Journal* 80 (8): 133.
Vilela, P. M., and D .Thompson (1999). 'Viscoelasticity: Why Plastic Bags Give Way When You Are Halfway Home.' *European Journal of Physics* 20 (1): 15–20.

CRISP SOUNDS

Crispness is associated with crunchiness, but your ears make a difference. That's the takeaway-and-chew-on-it message of an Oxford University study entitled 'The Role of Auditory Cues in Modulating the Perceived Crispness and Staleness of Potato Chips'.

THE ROLE OF AUDITORY CUES IN MODULATING THE
PERCEIVED CRISPNESS AND STALENESS OF POTATO CHIPS

MASSIMILIANO ZAMPINI[1] and CHARLES SPENCE[2]

Department of Experimental Psychology
University of Oxford, South Parks Road

The authors, experimental psychologists Massimiliano Zampini and Charles Spence, wax distantly poetical: 'We investigated whether the perception of the crispness and staleness of potato chips can be affected by modifying the sounds produced during the biting action. Participants in our study bit into potato chips with their front teeth while rating either their crispness or freshness using a computer-based visual analog scale.'

They recruited volunteers who were willing to chew, in a highly regulated way, on Pringles potato crisps. Pringles themselves are, as enthusiasts well know, highly regulated. Each crisp is of nearly identical shape, size, and texture, having been carefully manufactured from reconstituted potato goo.

The volunteers were unaware of the true nature of their encounter – that they would be hearing adulterated crunch sounds. But whatever risks this entailed were small. The experiment, Zampini and Spence take pains to say in their report, 'was performed in accordance with the ethical standards laid down in the 1964 Declaration of Helsinki. Participants were paid £5 for taking part in the study.'

Each volunteer sat in a soundproofed experimental booth, wearing headphones, facing a microphone, and operating a pair of foot pedals.

The headphones delivered Pringles crunch sounds that, though born in the chewer's mouth, had been captured by the microphone and electronically cooked. At times, the crunch sounds were delivered to the headphones with exacting, lifelike fidelity. At other times, the sounds were magnified. At still other times, only the high frequencies of the crunch were intensified.

The foot pedals were the means by which a volunteer could register his or her judgements as to (a) the crispness and (b) the freshness of a particular crisp.

Each crisp's crispness was judged from a single, headphone-enhanced bite delivered with the front teeth. Zampini and Spence adopted this approach for two reasons. It maximized the uniformity of the participant's contact with each crisp. And previous research, by others, showed that the sound of the first bite is what counts most for judging crispness.

The results? As the report puts it: 'The potato chips were perceived as being both crisper and fresher when either the overall sound level was increased, or when just the high frequency sounds (in the range of 2 kilohertz–20 kilohertz) were selectively amplified.'

Zampini and Spence say this gives new insight on an old research finding. In 1958, in the *Journal of Applied Psychology*, G. L. Brown 'reported that bread was judged as being fresher when wrapped in cellophane than when wrapped in wax paper'. The sound made by wrappers, they hazard, may have unappreciated influence.

There exists a Dutch study showing that generally you *can* judge a book by its cover. I mention it here only for contrast, because the Oxford report implies that maybe you can't judge the crunch of a crisp by the crackle of its wrapper.

In 2008, Zampini and Spence were awarded the Ig Nobel Prize in the field of nutrition for their efforts to electronically modify crisp sounds to seem crisper and fresher.

Zampini, Massimiliano, and Charles Spence (2004). 'The Role of Auditory Cues in Modulating the Perceived Crispness and Staleness of Potato Chips.' *Journal of Sensory Studies* 19 (5): 347–63.
Brown, G. L. (1958). 'Wrapper Influence on the Perception of Freshness in Bread.' *Journal of Applied Psychology* 42: 257–60.
Piters, Ronald A. M. P., and Mia J. W. Stokmans (2000). 'Genre Categorization and Its Effect on Preference for Fiction Books.' *Empirical Studies of the Arts* 18 (2): 159–66.

ENOUGH ALREADY

'Had enough?' This simple query drives Brian Wansink of Cornell University, in New York, to conduct experiment after experiment after experiment. Had enough popcorn? Had enough candy? Had enough rum and Coke? Wansink wants to know.

Most of the other experts on 'Had enough?' are nutritionists, mothers, or waiters. They serve up their conclusions in a sandwich of nutritionist, maternal, or waiterly intuition. Professor Wansink is an economist. He presents his thoughts atop beds of freshly harvested data.

Wansink methodically chews at the riddle of what makes a trencherman. He proceeds substance by substance.

As if calibrating his equipment, Wansink began with plain, pure water in bottles, publishing a paper in 1996 called 'Can Package Size Accelerate Usage Volume?' The answer, he says, is yes.

Five years later, Wansink and a colleague, the evocatively named graduate student Se-Bum Park, published a report in the journal *Food Quality and Preference*. It describes the experiment they conducted with patrons at a screening of the film *Payback*, which stars Mel Gibson. These discriminating cineastes munched free popcorn. The researchers noted that 'moviegoers who had rated the popcorn as tasting relatively unfavorable ate 61% more popcorn if randomly given a large container than a smaller one'.

The next year, 2002, saw publication of 'How Visibility and Convenience Influence Candy Consumption' in the journal *Appetite*. These candy results were as startling as the earlier popcorn findings. People ate more chocolate drops if the candy jar was kept on a desk rather than somewhere less handy or visible.

A study published in 2004 described a series of relatively complex experiments involving jelly beans and M&Ms. This was an attempt to probe how 'the structure of an assortment (e.g., organization and symmetry or entropy) moderates the effect of actual variety on perceived variety'.

The jelly beans caused problems. The report says merely that ' 23 [people] indicated that they did not like jelly beans and were dropped from the study. Five others were deleted from the analysis because they accidentally spilled the jelly beans or emptied the entire tray onto the table and scooped the jelly beans into their pockets.'

Wansink holds the university's John S. Dyson chair of marketing and applied economics. The chair was endowed by Robert R. Dyson to honour his brother. John S. Dyson created the 'I ♥ NY' tourism campaign, which, for more than three decades now, has not stopped serving up television ads, and magazine ads, and other ads, ad nauseum, to drive the phrase 'I ♥ NY' through the eyes and ears and into the brains of billions of human beings round the globe.

Wansink ascended the John S. Dyson chair in 2005, after spending eight years at the University of Illinois, where he held several titles, including that of Julian Simon memorial faculty fellow in marketing. That fellowship was endowed to honour the memory of Julian Simon, an economist who himself had an obsession with 'Had enough?' Simon's former colleagues say he was 'a man who won an international reputation for his buoyant and often controversial views on the limitless potential of human beings to meet and overcome the challenge of declining resources – views that won him the one-word description of "doomslayer".'

In Wansink's later experiments, people slurped rum and Cokes from glasses that were tall and slim or short and stout; munched roasted nuts and a pretzel variety mix from party bowls of various sizes; and spooned soup from bottomless bowls – a challenge of rising resources. The bowls were not literally bottomless – rather, they 'slowly and imperceptibly refilled as their contents were consumed'. A bottomless soup bowl experiment, in which people showed themselves to be nearly insatiable, earned Professor Wansink an Ig Nobel Prize, awarded in 2007, in the field of nutrition.

Wansink has revisited the popcorn question, and more recently, in the *BMJ*, the booze.

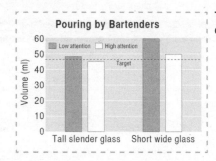

The long and short of bartenders' pours

Fans can hope that one day he will revisit the soup. In a press release some years ago, Wansink said, 'We thought it would be interesting to examine personality types based on strongly expressed soup preferences.' However, he has yet to publish on this topic in a peer-reviewed scientific journal.

Wansink, Brian (1996). 'Can Package Size Accelerate Usage Volume.' *Journal of Marketing* 60: 1–14.
— (2002). 'Changing Eating Habits on the Home Front: Lost Lessons from World War II Research.' *Journal of Public Policy & Marketing* 21 (1): 90–99.
Kahn, Barbara E., and Brian Wansink, (2004). 'The Influence of Assortment Structure on Perceived Variety and Consumption Quantities.' *Journal of Consumer Research* 30: 519–33.
Wansink, Brian, and Koert van Ittersum (2003). 'Bottoms Up! The Influence of Elongation on Pouring and Consumption Volume.' *Journal of Consumer Research* 30: 455–63.
— (2005). 'Shape of Glass and Amount of Alcohol Poured: Comparative Study of Effect of Practice and Concentration.' *BMJ* 331: 1512–14.
Wansink, Brian, and Junyong Kim (2005). 'Bad Popcorn in Big Buckets: Portion Size Can Influence Intake as Much as Taste.' *Journal of Nutrition Education and Behavior* 37 (5): 242–45.
Wansink, Brian, and Se-Bum Park (2000). 'Accounting for Taste: Prototypes that Predict Preference.' *Journal of Database Marketing* 7: 308–20.
— (2001). 'At the Movies: How External Cues and Perceived Taste Impact Consumption Volume.' *Food Quality and Preference* 12 (1): 69–74.
Wansink, Brian, James E. Painter, and Jill North (2005). 'Bottomless Bowls: Why Visual Cues of Portion Size May Influence Intake.' *Obesity Research* 13 (1): 93–100.
Painter, James E., Brian Wansink, and Julie B. Hieggelke (2002). 'How Visibility and Convenience Influence Candy Consumption.' *Appetite* 38 (3): 237–38.
Wansink, Brian, and Matthew M. Cheney (2005). 'Serving Bowls, Serving Size, and Food Consumption: A Randomized Controlled Trial.' *JAMA – Journal of the American Medical Association* 293 (14): 1727–28.

SIX

MONEY CAN BE VALUABLE

IN BRIEF

'HOW HIGH CAN A DEAD CAT BOUNCE?': METAPHOR AND THE HONG KONG STOCK MARKET'
by Geoff P. Smith (published in *Linguistics and Language Teaching*, 1995)

Some of what's in this chapter: Money destruction in your skull • The lure of piracy, for economists • 2127 rounds of rock, paper, scissors per day • A careful look inside Russian underwear • Foucault and footie • The shapely heads of CEOs • Griffiths, slot-machine psychologist • Perfume for the poor • Author, author, author, author, author, author, author, author, author, author • $100,000,000,000,000 • Corporate tiers of a clown

ALL TORN UP

If you have never watched someone rip up large amounts of cash, you may be unsure as to how the different parts of your brain would respond in the event that you did see someone tearing valuable banknotes into tiny, worthless shreds. A new study may help you predict what would happen.

The study is called 'How the Brain Responds to the Destruction of Money'. It tells how the brains of twenty Danish persons, all of them adults with no history of psychiatric or neurological disease, responded as they watched videos of somebody destroying lots of Danish money.

If you are not Danish, you might now expect that your brain would respond in rather the same way, were this to involve your own native currency (pounds, euros, dollars, or whatever). The study – performed by Uta Frith and Chris Frith of University College London, together with Joshua Skewes, Torben Lund, and Andreas Roepstorff of Aarhus University, Denmark, and Cristina Becchio of the University of Turin, Italy – makes no specific claims for non-Danish brains or money, however.

Here, in the scientists' words, is what the volunteers saw: 'A series of videos in which different actions were performed on actual bank- notes with a value of either 100 Kroner (approximately 13 euro/18 US dollar) or 500 Kroner (approximately 67 euro/91 US dollar), or on valueless pieces of paper of the same size ... We contrasted actions that were appropriate to money (folding or looking at valuable notes or valueless paper) and actions that were inappropriate (tearing or cutting notes or paper).'

The Danes had their heads inside a functional magnetic resonance imaging scanner (fMRI), which recorded some of their brain activity. The researchers also asked each volunteer some questions, including 'How did it make you feel?' All of this, says the document, 'confirmed that participants felt less comfortable during observation of destroying actions performed on money'. An additional finding: the volunteers felt more 'aroused' when watching anything happen to money than when watching the same things happen to worthless paper.

The scientists find the brain scans to be especially interesting. The activity patterns, they say, are similar to something they've seen before: the 'use of concrete tools, such as hammers or screwdrivers, has been associated with activation of a left hemisphere network including the posterior temporal cortex, supramarginal gyrus, inferior parietal lobule, and lateral precuneus. Here we demonstrate that observing bank notes being cut up or torn, a critical violation of their function, elicits activation within the same temporo-parietal network. Moreover, this activation is the greater the higher the value of the banknote.'

They caution that the story must be more complex, that your brain probably regards money in several – differing – ways. They note, for example, the existence of published studies that 'suggest that money can also act as a drug'.

Team member Chris Frith, by the way, was part of a group that gathered evidence about the brains of London taxi drivers being more highly developed than those of their fellow citizens. That study's findings, honoured with the 2003 Ig Nobel Prize in medicine, appear to be unrelated to these later cautions.

Becchio, Cristina, Joshua Skewes, Torben E. Lund, Uta Frith, Chris Frith, and Andreas Roepstorff (2011). 'How the Brain Responds to the Destruction of Money.' *Journal of Neuroscience, Psychology, and Economics* 4 (1): 1–10.

Lea, Stephen E. G., and Paul Webley (2006). 'Money as Tool, Money as Drug: The Biological Psychology of a Strong Incentive.' *Behavioral and Brain Sciences* 29: 161–209.

Maguire, Eleanor, David Gadian, Ingrid Johnsrude, Catriona Good, John Ashburner, Richard Frackowiak, and Christopher Frith (2000). 'Navigation-Related Structural Change in the Hippocampi of Taxi Drivers.' *Proceedings of the National Academy of Sciences* 97 (8): 4398–403.

THE INVISIBLE HOOK OF PIRATE ECONOMICS

Pirates are a practical lot, at least in theory. The theory was supplied in 2007 by Peter T. Leeson, an assistant professor of economics at West Virginia University. He is of the opinion that pirates pioneered some basic economics.

In a study called 'Pi*rational* Choice: The Economics of Infamous Pirate Practices', Leeson 'investigates the internal governance institutions of violent criminal enterprise by examining the law, economics, and organization of pirates'. These were the classical pirates of the seventeenth and eighteenth century, especially those who practised professionally in and around the West Indies and in the waters around Madagascar. Leeson's study appeared before chestsful of information began to surface in 2008 and 2009 in Wall Street, in the City of London, and at other romantic places where peril and opportunity drive many a captain of finance to pursue plunder.

'Pirate governance created sufficient order and cooperation to make pirates one of the most sophisticated and successful criminal organizations in history', writes Leeson. 'To effectively organize their banditry, pirates required mechanisms to prevent internal predation, minimize crew conflict, and maximize piratical profit.'

Pirates, he argues, invented a system of checks and balances 'to constrain captain predation', and devised democratic constitutions to 'create law and order' among themselves. 'Remarkably,' points out Leeson, 'pirates adopted both of these institutions before the United States or England.'

These pirate practices of the past now read like a 'best practices' primer on economics and finance. Successful buccaneers learned how to manage organizational growth: 'Many pirate crews were too large to fit in one ship. In this case they formed pirate squadrons ... Multiple pirate ships often joined for concerted plundering expeditions. The resulting pirate fleets could be massive.' They recognized that the big pirates had to be restrained from completely plundering the treasures of the little pirates under their command. Leeson uses simplified mathematical models to explain how this was achieved. 'Consider a pirate ship of complete but imperfect information with a captain and two "factions" of ordinary pirates that together comprise the ship's crew', he says. 'The captain moves first and decides whether to prey on the crew or not. If he preys on both factions simultaneously, they join together to overthrow him, so this is not an option he entertains. He can only prey on one faction at a time.'

The study works out the theoretical consequences, and summarizes them in two graphs captioned 'The Threat of Captain Predation' and 'Piratical Checks and Balances: Constraining Captain Predation'. Examining the lines that connect the various nodes, one can follow the workaday machinations of pirate economic life, and see how these resolve into multiple equilibria and a collection of expected 'payoffs'.

---

2   Buried Treasure: A Note on Sources

To explore the economics of infamous pirate practices, I consider late 17th- and early 18th-century (1660-1730) sea bandits who occupied the waterways that formed major trade routes surrounding the Bahamas, connecting Europe and the North American sea coast, between Cuba and Haiti, and around Madagascar. These areas encompass major portions of the Atlantic Ocean, Indian Ocean, Caribbean Sea, and Gulf of Mexico. The trade routes connecting the Caribbean, North America's

---

A note on sources from 'Pi*rational* **Choice'**

One sees at a glance how organizations of pirates come (in Leeson's theory) to restrain or deflect themselves from destroying their own organizations. They enable themselves to keep working for the greater, more effective plunder of the larger, non-pirate community.

Leeson, Peter T. (2010). 'Pi*rational* Choice: The Economics of Infamous Pirate Practices.' *Journal of Economic Behavior and Organization* 76 (3): 497–510.

— (2007). 'Trading with Bandits.' *Journal of Law and Economics* 50 (2): 303–21.

— (2009). 'The Invisible Hook: The Law and Economics of Pirate Tolerance.' *New York University Journal of Law and Liberty* 4: 139–71.

—, and C. Coyne. 'The Economics of Computer Hacking.' *Journal of Law, Economics and Policy* 1(2) 2006: 511–32.

MAY WE RECOMMEND

'HOW SMART ARE THE SMART GUYS? A UNIQUE VIEW FROM HEDGE FUND STOCK HOLDINGS'

by John M. Griffin and Jin Xu (published in the *Review of Financial Studies*, 2009)

The authors, at the University of Texas at Austin and Zebra Capital Management, report: 'We provide the first comprehensive examination of hedge funds' long-equity positions and the performance of these stock holdings ... Overall, our study raises serious questions about the proficiency of hedge fund managers.'

ROCK, PAPER, MONKEYS

Among scholars of the game of rock paper scissors, only a tiny minority also study monkeys. This fact, by itself, may explain why no studies were published until 2005 about what happens when monkeys play rock-paper-scissors.

Daeyeol Lee, Benjamin P. McGreevy, and Dominic J. Barraclough of the University of Rochester, in New York, wrote that first, and so far the only, report on the subject: 'Learning and Decision Making in Monkeys during a Rock-Paper-Scissors Game'. Lee, the lead author, has since moved to Yale University, where he is an associate professor of neurobiology.

The test subjects were male rhesus monkeys. No one explained to them the rules of the game: that rock breaks scissors, scissors cut paper, paper covers rock. The scientists wanted to see whether and how the monkeys would learn from the brute experience of playing game after game after game. Whenever a monkey did well, it got a sudden, sweet reward: a little drop of juice after each tie, two drops after each win. It received nothing – not even faint boos – after a loss.

For reasons unstated, the scientists chose not to use real rocks, paper, or scissors. Instead, they rigged a computer to display crude patterns of dots and circles. Different patterns represented a rock, a sheet of paper, and a pair of scissors. The monkeys were not informed as to which symbol stood for what object.

The monkeys were treated in a manner that is familiar to many hard-core computer gamers. Each sat in a chair, facing a monitor that flashed the symbols for rock, paper, and scissors. Rather than being asked to make the traditional physical gesture for rock, paper, or scissors, the monkey was expected to cast its gaze towards the symbol of its choice. The scientists tracked each monkey's eye movements, using a German-manufactured Thomas-ET49 high-speed video-based eye-tracking machine, and digitally recorded the entire sequence of each monkey's choices.

There were only two monkeys. They worked hard.

One monkey played rock-paper-scissors for forty-one days, making a total of 87,200 choices, an average of 2127 rounds every day. The other monkey played for fifty-two days, making a total of 82,661 choices, an average of 1589 rounds per day. Over the long haul, each monkey chose paper about as often as it picked scissors. Both monkeys displayed a slightly irrational aversion to rocks.

The scientists used economics theory to critique the overall performance, saying: 'Each animal displayed an idiosyncratic pattern substantially deviating from Nash equilibrium.' The Nash equilibrium was conceived by John Forbes Nash, who was awarded a Nobel Prize in 1994 for his 'pioneering analysis of equilibria in the theory of non-cooperative games'. He was also the subject of the 2001 film *A Beautiful Mind*.

Because only two monkeys were tested, Daeyeol et al. admitted that it 'was difficult to conclude' exactly which strategies a monkey uses to play the game. 'This', they write, 'remains to be investigated in future studies.'

Lee, Daeyeol, Benjamin P. McGreevy, and Dominic J. Barraclough (2005). 'Learning and Decision Making in Monkeys During a Rock-Paper-Scissors Game.' *Cognitive Brain Research* 25 (2): 416–30.

SOVIET UNDERWEAR READ

Olga Gurova studies the cultural history of underwear in the Soviet Union. 'When I am talking about Soviet underwear', she says, 'I mean the underwear that appeared after the 1917 Revolution.'

Gurova is based at the Academy of Finland department of social research. In 2005–06, she spent a year in the US as a Fulbright fellow, and her public lectures helped to fill the information gap that developed during the Cold War.

In the 1920s, Soviet magazines touted a 'regime of cleanliness' for the proletariat. 'Underwear', explains Gurova, 'was a compulsory part of that regime.' A goal was established: everyone should have at least two sets of underwear, and should change sets at least once every seven to ten days. Mass production was cranked up, underclothing the populace in officially healthy, comfortable, hygienic long johns, boxers, undershirts, and bras. Gurova's research shows that most of these items were 'spacious', and that 'there was no big difference in design between male and female underclothes'.

Having pored over masses of documentation, Gurova infers that during the 1920s 'Soviet underwear was not about sex, it was about sport'. Sports outfits – T-shirts, shorts, and sleeveless shirts – became the basic prototypes. Petticoats, seen as bulky and old-fashioned, faded from the scene, as did corsets. Underwear design quickly adapted to better serve Soviet women's wide-ranging physical activities in the factory and the kitchen. In contrast to most European countries, reports Gurova, 'the Soviet revolution canceled corsets and dressed women in bras more quickly'.

Gurova hypothesizes that, after the 1920s, there were three major periods in the history of Soviet underwear. The 1930s and 1940s were characterized by a Joseph Stalin speech, in 1935, proclaiming that Soviet life was becoming more abundant and joyful. Women's underwear became somewhat feminine. For both sexes, undergarments could now be in certain colours. According to Gurova: 'If previously they were white in color, according to hygienic reasons, later they become black, vinous, khaki or dark blue, and the explanation was the opposite than previous: dark colors become dirty slower.'

In the 1950s and 1960s, Premier Nikita Khrushchev increased Soviet interaction with other countries. Clothing styles were on Soviet minds. Soviet stores offered a wider, if not quite dizzying, array of consumer items. Soviet underwear became 'a means of personal expression'.

The final period, the 1970s and 1980s, was marked by consumer goods shortages – and by a government campaign against obesity, with the slogan 'To be plump is no good'. For many citizens, Gurova says, 'it was hardly possible to buy undergarments that fitted well'.

It was here that the Soviet peoples showed their resilience. Gurova says that 'manipulations with clothes at home became very popular: people sewed clothes, repaired them, and constructed new clothes from the old ones ... The Soviet man overcame the shortage, personified and privatized those standard clothes.'

These are the barest facts. Dr Gurova plans to cover them more fully with a book.

Gurova, Olga (2005). 'Making of the Body: Cultural History of Underwear in Soviet Russia.' Paper presented at the Russian, East European, and Eurasian Center, University of Illinois at Urbana-Champaign, 29 November.

IN BRIEF

'HOW HELLO KITTY COMMODIFIES THE CUTE, COOL, AND CAMP: "CONSUMUTOPIA" VERSUS "CONTROL" IN JAPAN'
by Brian J. McVeigh (published in the *Journal of Material Culture*, 2000)

Evidence: Hello Kitty 'bankbook and bank card'

FOUCAULT ON MANAGEMENT

Of all the football leagues for all the players in the world, the Australian Football League is the first to sponsor research that overtly applies the work of the French philosopher Michel Foucault.

Australian football is not soccer, nor is it American football. It is, fans and players like to point out, based on a different philosophy from anything else that answers to the name 'football'. The Australian game even answers to its own, special nickname: footy. Australian behaviourists Peter Kelly and Christopher Hickey elucidate one aspect of the game's philosophy in a study they call 'Foucault Goes to the Footy: Professionalism, Performance, Prudentialism and Playstations in the Life of AFL Footballers'.

They went public with their work in 2004, when both were based at Australia's Deakin University. Kelly has since moved to Monash University. He is also an honorary senior fellow at the University of Hull, UK. Both men are conversant with the thoughts of Michel Foucault, the bespectacled, bald philosopher who died in 1984, the year the Essendon Bombers won the footy championship, coming from four goals behind at the three-quarter mark of the ultimate game to decimate the defending champions, Hawthorn Hawks.

Foucault famously said, 'Madness, death, sexuality, crime; these are the subjects that attract most of my attention.' Several million footy-mad Australians would say much the same, be they supporters of Geelong, St Kilda, Adelaide, Carlton, Collingwood, or any of the eleven other clubs in the Australian Football League.

Kelly and Hickey say their research is 'informed by Foucault's later work on the care of the Self to focus on the ways in which player identities are governed by coaches, club officials, player agents and the AFL Commission/Executive; and the manner in which players conduct themselves in ways that can be characterised as professional – or not'.

That is a mouthful. It comes down to using Foucault's philosophical ideas to help footy clubs choose players who will be worth the clubs' substantial recruitment and salary investment.

The ideas are drawn primarily from Foucault's 'The Ethics of the Concern for Self as a Practice of Freedom' and 'Subjectivity and Truth', essays he wrote late in life, during the period when the Australian Football League was still calling itself the Victorian Football League.

Then, player salaries were lower and footy clubs were almost carefree in their risk-management practices. Nowadays, the write-off cost of a defective or disruptive footy player is bigger, and thus so are the worries of prudent footy executives. Foucault helps them tackle those worries.

Elsewhere, the business community has been, on the whole, slow to adopt Foucault's contributions to the philosophy of accounting. But the Australian Football League has built itself a platypus of a game by incorporating odd elements from the most unexpected places. It is unafraid to throw something different – even a dead French intellectual icon – into its business plans.

Kelly, Peter, and Christopher Hickey (2004). 'Foucault Goes to the Footy: Professionalism, Performance, Prudentialism and Playstations in the Life of AFL Footballers.' Paper presented at the TASA Annual Conference, Latrobe University, December.

A GOOD HEAD SHAPE FOR BUSINESS

A new line of American-British research suggests that the shape of a chief executive officer's head can indicate how well his firm will prosper. The shape also predicts whether or not the chief executive will act immorally.

The research offers a mathematical tool that financial analysts can add to their professional kit bag: the chief executive officer's facial width-to-height ratio. The 'chief executive facial WHR', for short. The research and its financial implications are outlined in a study called 'A Face Only an Investor Could Love: CEOs' Facial Structure Predicts Their Firms' Financial Performance', to be published in the journal *Psychological Science*.

Research Report

A Face Only an Investor Could Love: CEOs' Facial Structure Predicts Their Firms' Financial Performance

Elaine M. Wong[1], Margaret E. Ormiston[2], and Michael P. Haselhuhn[3]
[1]Department of Communication, University of Wisconsin–Milwaukee; [2]Department of Organisational Behaviour, London Business School; and [3]Sheldon B. Lubar School of Business, University of Wisconsin–Milwaukee

The authors, Elaine Wong and Michael Haselhuhn at the University of Wisconsin-Milwaukee, and Margaret Ormiston at London Business School, explain the significance of their work. Prior researchers, they say, failed 'to empirically identify physical traits that predict leadership success' or predict 'the ability of leaders to achieve organizational goals'.

Their discovery, in their view, constitutes a breakthrough: 'We identify a specific physical trait, facial structure, of leaders that correlates with organizational performance. Specifically, chief executive officers with wider faces (relative to facial height) achieve superior firm financial performance.'

The story is not always that simple, the researchers caution, nor is it guaranteed: 'The relationship between chief executive facial structure and financial performance is moderated by the decision-making dynamics of the leadership team.'

Wong, Haselhuhn, and Ormiston painstakingly examined the financial performance and chief executive facial measurements of General Electric, Hewlett Packard, Nike, and fifty two other publicly traded Fortune 500 firms for the period from 1996 through 2002. The companies are big, averaging $38 billion in annual sales and about 120,000 employees.

The researchers obtained chief executive facial photos from the Internet, using these as raw data from which to calculate each chief executive facial WHR. They looked up each firm's return on assets (in financial industry shorthand, the ROA), using that as the measure of the company's financial performance.

Wong and Haselhuhn spell out their logic in a study called 'Bad to the Bone: Facial Structure Predicts Unethical Behaviour', published in the *Proceedings of the Royal Society B* in 2011. They tell of experiments conducted on students, which showed 'that men with wider faces (relative to facial height) are more likely to explicitly deceive their counterparts in a negotiation, and are more willing to cheat in order to increase their financial gain'.

They explain the mechanism that might make this happen. Prior research indicated that wide-faced men are 'associated with more aggressive behaviour'. If 'observers respond to facial cues by deferring to men whom they perceive to be aggressive based on their facial WHR, these men may find it easier to take advantage of others. Similarly, if men with greater facial WHRs are treated in ways that make them feel more powerful, this may foster a psychological sense of power, which then affects ethical judgement and behaviour.'

The facial indicators, say the researchers, are more reliable in men than in women.

The American Psychological Association, which is publishing the CEO study, issued a press release that finishes with this warning: 'Don't run out and invest in wide-faced CEOs' companies, though.

Wong and her colleagues also found that the way the top management team thinks, as reflected in their writings, can get in the way of this effect. Teams that take a simplistic view of the world, in which everything is black and white, are thought to be more deferential to authority; in these companies, the CEO's face shape is more important. It's less important in companies where the top managers see the world more in shades of gray.'

Wong's work reinforces an ancient and reluctantly treasured belief: that having a good head on your shoulders does not, by itself, ensure success.

Wong, Elaine M., Margaret E. Ormiston, and Michael P. Haselhuhn (2011). 'A Face Only an Investor Could Love: CEOs' Facial Structure Predicts Their Firms' Financial Performance.' *Psychological Science* 22 (12): 1478–83.
Haselhuhn, Michael P., and Elaine M. Wong (2011). 'Bad to the Bone: Facial Structure Predicts Unethical Behaviour.' *Proceedings of the Royal Society B* online, http://rspb.royalsocietypublishing.org/content/early/2011/06/29/rspb.2011.1193.

SOME PSYCHOLOGY OF THE FRUIT MACHINE

It's hard to get good payoffs from slot machines, yes. But it's also hard to get good information from slot-machine gamblers, and that made things awkward for British psychologists Mark Griffiths of Nottingham Trent University and Jonathan Parke of Salford University. They explained the issue in a monograph entitled 'Slot Machine Gamblers: Why Are They So Hard to Study?'

Griffiths and Parke published their report in the *Electronic Journal of Gambling Issues*. 'We have both spent over 10 years playing in and researching this area', they wrote, 'and we can offer some explanations on why it is so hard to gather reliable and valid data.'

Here are three from their long list.

> FIRST, *gamblers become engrossed in gambling.* 'We have observed that many gamblers will often miss meals and even utilise devices (such as catheters) so that they do not have to take toilet breaks. Given these observations, there is sometimes little chance that we as researchers can persuade them to participate in research studies.'
>
> SECOND, *gamblers like their privacy.* They 'may be dishonest about the extent of their gambling activities to researchers as

well as to those close to them. This obviously has implications for the reliability and validity of any data collected.'

THIRD, *gamblers sometimes notice when a person is spying on them*. 'The most important aspect of non-participant observation research while monitoring fruit machine players is the art of being inconspicuous. If the researcher fails to blend in, then slot machine gamblers soon realise they are being watched and are therefore highly likely to change their behavior.'

Griffiths is one of the world's most published scholars on matters relating to the psychology of slot-machine gamblers, with at least twenty-seven papers that mention so-called fruit machines, so-called for their bounty of cherries, oranges, and other juicy prizes. (It's helpful to note that, in the UK, games of pure chance are not allowed, and so fruit machines require some element of 'skill'; Griffiths and Parke don't seem to obsess about the varied nomenclature.)

Griffith's titles range from 1994's appreciative 'Beating the Fruit Machine: Systems and Ploys Both Legal and Illegal' to 1998's admonitory 'Fruit Machine Gambling and Criminal Behaviour: Issues for the Judiciary'. Women get special attention ('Fruit Machine Addiction in Females: A Case Study'), as do youths ('Adolescent Gambling on Fruit Machines' and several other monographs). There is the humanist perspective ('Observing the Social World of Fruit-machine Playing') as well as that of the biomedical specialist ('The Psychobiology of the Near Miss in Fruit Machine Gambling'). The *International Journal of Mental Health and Addiction* ran a paean from a researcher who said: 'In the problem gambling field we don't exhibit the same adulation as music fans for their idols but we have our superstars and for me, Mark Griffiths is one.'

Griffiths and Parke collaborate often. (Strangers to their work might wish to begin by reading 'The Psychology of the Fruit Machine'.) Their fruitful publication record reminds every scholar that, even when a subject is difficult to study, persistence and determination can yield a rewarding payoff.

Parke, Jonathan, and Mark Griffiths (2002). 'Slot Machine Gamblers: Why Are They So Hard to Study?' *eGambling: Electronic Journal of Gambling Issues* 6.
McKay, Christine (2007). 'A Luminary in the Problem Gambling Field: Mark Griffiths.' *International Journal of Mental Health and Addiction* 5 (2): 117–22.

Griffiths, Mark. (1994). 'Beating the Fruit Machine: Systems and Ploys Both Legal and Illegal.' *Journal of Gambling Studies* 10: 287–92.

Griffiths, Mark, and Paul Sparrow (1998). 'Fruit Machine Gambling and Criminal Behaviour: Issues for the Judiciary.' *Justice of the Peace* 162: 736–39.

Griffiths, Mark. (2003). 'Fruit Machine Addiction in Females: A Case Study.' *eGambling: Electronic Journal of Gambling Issues* 8.

— (1996). 'Adolescent Gambling on Fruit Machines.' *Young Minds Magazine* 27: 10–11.

— (1996). 'Observing the Social World of Fruit-machine Playing.' *Sociology Review* 6 (1): 17–18.

— (1991). 'The Psychobiology of the Near Miss in Fruit Machine Gambling.' *Journal of Psychology* 125: 347–57.

—, and Jonathan Parke (2003). 'The Psychology of the Fruit Machine.' *Psychology Review* 9 (4): 12–16.

MAY WE RECOMMEND

'PHYSIOLOGICAL AROUSAL AND SENSATION-SEEKING IN FEMALE FRUIT MACHINE GAMBLERS'

by K. R. Coventry and B. Constable (published in *Addiction*, 1999)

The authors, who are at the University of Plymouth, UK, conclude: 'Gambling alone is not enough to induce increases in heart rate levels for female fruit machine gamblers; the experience of winning or the anticipation of that experience is necessary to increase heart rate levels.'

THE VALUE OF PERFUME FOR THE POOR

What do destitute people have in mind when they haggle for famous-name perfume? Luuk van Kempen attacked the question head-on. He describes his experiment, and his thinking, in a report called 'Are the Poor Willing to Pay a Premium for Designer Labels? A Field Experiment in Bolivia'.

Van Kempen, who is based at Tilburg University in the Netherlands, is trying to get at a deeper question: 'Why do the poor buy status-intensive goods, while they suffer from inadequate levels of basic needs satisfaction?' His study, which appears in the journal *Oxford Development Studies*, proceeds in social-scientific fashion, listing each of his assumptions, and shaking each one to see whether it is true.

FIRST QUESTION: Will impoverished Bolivians bargain for designer-label perfume? To find out, van Kempen had them play a what-if game known as a Becker-DeGroot-

Marschak elicitation scheme. 'Even subjects who have received little formal education', he explains, 'should be able to understand the procedure.' Indeed, 104 residents of a poor neighbourhood in the city of Cochabamba understood the procedure well enough to quarrel with van Kempen over the price of perfumes.

SECOND QUESTION: How does one know this neighbourhood was poor? From the lack of safe water and of good sanitation facilities. The facilities 'often consisted of one single latrine, were mostly shared among seven to 10 families'. Van Kempen explains that, for academic purposes, this had its benefits: 'The experiment provided an implicit test of Maslow's hierarchy-of-needs [theory], which suggests that people do not indulge in symbolic consumption, i.e. the acquisition of goods that satisfy belongingness and status needs, as long as "basic needs" are not satisfied.'

THIRD QUESTION: Is it reasonable to assume that a lack of access to safe water and sanitation truly indicates poverty? Yes, van Kempen concludes, citing a 2001 study of a different part of the city, which also lacked water and sewers. In that neighbourhood, eighty-seven percent of the population had incomes averaging about £1, or $1.80, a month. 'Hence', he says, 'lack of access to basic services is a reasonably good proxy for income poverty.'

The main part of the study is titled 'The Logo Premium: Do the Poor See Beyond Their Nose?' This is where van Kempen describes his experiment. It involved bottles of perfume, some with a Calvin Klein label, others without. The perfume inside all the bottles smelled – and was – the same. Why Calvin Klein? 'Because it is one of the best known designer brands in Bolivia.'

Each poor person got to choose which perfume to buy – the Calvin Klein or the generic alternative – and bargain over the price she or he would pay for one versus the other.

About forty percent of these extremely low-income Bolivians were willing to pay extra for the designer name. What, statistically, was in their minds? Social caché, says van Kempen, the ability to walk with their noses in the air, regardless of what they might smell there.

Van Kempen, Luuk (2004). 'Are the Poor Willing to Pay a Premium for Designer Labels? A Field Experiment in Bolivia.' *Oxford Development Studies* 32 (2): 205–24.

EXTREME SPEED WRITING

Philip M. Parker is the world's fastest book author, and given that he had been at it for only five years or so when I contacted him in 2008, and already had more than 85,000 books to his name, he is likely the most prolific, as well as the most titled.

Parker is also the most wide-ranging of authors – the phrase 'shoes and ships and sealing wax, cabbages and kings' is not half a percent of it. Nor are these particular subjects foreign to him. He has authored some 188 books related to shoes, ten about ships, 219 books about wax, six about sour red cabbage pickles, and six about royal jelly supplements.

To begin somewhere, let's note that Parker is the author of the book *The 2007–2012 Outlook for Bathroom Toilet Brushes and Holders in the United States*, which is 677 pages long, sells for £250/$495, and is described by the publisher as a 'study [that] covers the latent demand outlook for bathroom toilet brushes and holders across the states and cities of the United States'. (A later edition, covering 2009–2014, retails for £495/$795. Further Parkerian volumes and updated pricing can be expected to appear automatically in the years, decades, and centuries beyond.)

Here's a minuscule (compared to the entire, ever-growing list) sampling of Philip M. Parker titles:

> *The 2007–2012 World Outlook for Rotary Pumps with Designed Pressure of 100 P.s.i. or Less and Designed Capacity of 10 G.p.m. or Less*
>
> *Avocados: A Medical Dictionary, Bibliography, and Annotated Research Guide*
>
> *Webster's English to Romanian Crossword Puzzles: Level 2*
>
> *The 2007–2012 Outlook for Golf Bags in India*
>
> *The 2007–2012 Outlook for Chinese Prawn Crackers in Japan*
>
> *The 2002 Official Patient's Sourcebook on Cataract Surgery*
>
> *The 2007 Report on Wood Toilet Seats: World Market Segmentation by City*
>
> *The 2007–2012 Outlook for Frozen Asparagus in India*

Parker is a professor of management science at INSEAD, the international business school based in Fontainebleau, France. Professor Parker is no dilettante. When he turns to a new subject, he seizes and shakes it till several books, or several hundred, emerge. About the outlook for bathroom toilet brushes and holder, Parker has authored at least six books. There is his *The 2007–2012 Outlook for Bathroom Toilet Brushes and Holders in Japan*, and also *The 2007–2012 Outlook for Bathroom Toilet Brushes and Holders in Greater China*, and also *The 2007–2012 Outlook for Bathroom Toilet Brushes and Holders in India*, and also *The 2007 Report on Bathroom Toilet Brushes and Holders: World Market Segmentation by City*.

When I first encountered Parker's output, Amazon.com offered 85,761 books authored by him. Parker himself said the total was well over 200,000. The number was then and is (even as you read these words, whenever you read them, possibly even if Professor Parker has been gone for decades or centuries) probably still on the rise.

How is this all possible? How does one man do so much? And why?

Parker created the secret to his own success. He invented what he calls a 'method and apparatus for automated authoring and marketing' – a machine that writes books. He says it takes about twenty minutes to write one.

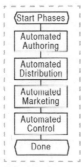

Fig. 1 of 13 from 'Method and Apparatus for Automated Authoring and Marketing' – 'the embodiment of the present invention'

Turn to page 16 of his patent, and you will see him answer the question, 'And why?'

Parker quotes a 1999 complaint, waged by *The Economist* magazine, that publishing 'has continued essentially unchanged since Gutenberg. Letters are still written, books bound, newspapers mostly printed and distributed much as they ever were.'

'Therefore', says Parker, 'there is a need for a method and apparatus for authoring, marketing, and/or distributing title materials automatically by a computer.' He explains that 'Further, there is a need for an automated system that eliminates or substantially reduces the costs associated with human labor, such as authors, editors, graphic artists, data analysts, translators, distributors, and marketing personnel.'

The book-writing machine works simply, at least in principle. First, one feeds it a recipe for writing a particular genre of book – a tome about crossword puzzles, say, or a market outlook for products, or maybe a patient's guide to medical maladies. Then hook the computer up to a big database full of info about crossword puzzles or market information or maladies. The computer uses the recipe to select data from the database and write and format it into book form.

Nothing but the title need actually exist until somebody places an order – typically via an online, automated bookseller. At that point, a computer assembles the book's content and prints up a single copy.

Among Parker's one hundred best-selling books (as ranked by Amazon) one finds surprises. His fifth-best seller in 2008 was *Webster's Albanian to English Crossword Puzzles: Level 1*. Bestseller No. 21: *The 2007 Import and Export Market for Seaweeds and Other Algae in France*. No. 66 is the aforementioned *The 2007–2012 Outlook for Chinese Prawn Crackers in Japan*. And rounding out the list, at No. 100, is *The 2007–2012 Outlook for Edible Tallow and Stearin Made in Slaughtering Plants in Greater China*.

Parker appears also to be enthusiastic about books authored the old-fashioned way. He has already written five of them.

Parker, Philip M. (2005). 'Method and Apparatus for Automatic Authoring and Marketing.' US Patent No. 7,266,767, 31 October.

THE HUNDRED TRILLION DOLLAR BOOK

Gideon Gono, author of the barnstorming book *Zimbabwe's Casino Economy – Extraordinary Measures for Extraordinary Challenges*, displays a rare, perhaps unique kind of scholarly reserve. He is a scholar, with a PhD from Atlantic International University, a mostly distance-learning institution based in the US, with a website that proclaims 'Atlantic International University is not accredited by an accrediting agency recognized by the United States Secretary of Education'. And he has reserve, or rather Reserve, with a capital 'R'. Since December 2003, Gideon Gono has been the governor of Zimbabwe's Reserve Bank. His term expires in 2013.

In 2009, Gono was awarded the Ig Nobel Prize in mathematics. The Ig Nobel citation lauds him for giving people a simple, everyday way to cope with a wide range of numbers – from the very small to the very big – by having his bank print bank notes with denominations ranging from one cent ($.01) to one hundred trillion dollars ($100,000,000,000,000).

During 2007 and 2008, Zimbabwe's inflation rate rose past Olympian heights: topping 231 million percent, by Gideon Gono's reckoning; and reaching 89,700,000,000,000,000,000,000 percent, according to a study done by Dr Steve H. Hanke of Johns Hopkins University, in Baltimore, Maryland, and the Cato Institute.

The book explains that every larger, richer country than Zimbabwe will face the same problems, at which time they will appreciate Gono's extraordinary skill at meeting such extraordinary challenges. Gono modestly shares the credit, writing on the very first page: 'I am especially indebted to my principal, President Robert Mugabe'.

Gono's talents were spotted by other influential persons. 'I was both humbled and surprised', he writes, 'to get an approach from [US] Ambassador [to Zimbabwe James] McGee on 25 July 2008 with an offer which he said was from President George W. Bush and Secretary Condoleeza Rice and the President of the World Bank for me to take a position in Washington as a Senior Vice President of the World Bank.'

He confides that later, 'my staff and I were amused to see the steady mushrooming of rather shameless news stories in some quarters of the Western Press and its allied media claiming that I had approached the United States authorities seeking their help to secure asylum for me and my family in some banana republic or that I somehow wanted to betray President Mugabe and Zimbabwe's national leadership and to run away from Zimbabwe in the face of what was alleged to be the collapse of the economy and President Mugabe's rule.'

Gono emphasizes the importance of sticking to one's principles. 'My team and I were guided by the philosophy', he writes, that 'where appropriate, short-term inflationary surges are a necessary cost to the achievement of medium to long-range growth in the economy.'

The book is, at heart, a 232-page literary fleshing-out of an eighteen-word statement issued by the Reserve Bank of Zimbabwe on 21 January 2008: 'Blaming the Government, the Reserve Bank or the Governor all the time is *unacceptable* and will be met with serious consequences.'

Gono, Gideon (2008). *Zimbabwe's Casino Economy – Extraordinary Measures for Extraordinary Challenges.* Harare: ZPH Publishers.

TRINKAUS ON TROLLEYS

Psychological Reports, 2004, 94, 1442-1443.   © Psychological Reports 2004

CLEARING THE SUPERMARKET SHOPPING CART: AN INFORMAL LOOK [1]

JOHN W. TRINKAUS

Zicklin School of Business
Baruch College, City University of New York

Summary.—An informal enquiry of the behavior of 500 supermarket shoppers clearing carts of litter prior to entering the store showed that 69% dumped the rubbish into another cart, 26% dropped it on the sidewalk, and 5% deposited it in a trash container.

Shopping carts are a window, however small, to our inner being.

'Some people entering supermarkets, to do their food shopping, seem to prefer to start their venture with a clean cart – one that is free of litter. However, many times a number of the available carts are not free of the leavings of previous shoppers, for example, store circulars, cash register receipts, shopping lists, plastic bags, produce remnants, facial tissues, and candy wrappers. Such folks, when finding that the cart at the end of the queue is not "clean", face a decision: push the cart to the side and try the next one, use it anyway, or somehow get rid of the material in the cart. It is this third alternative that was looked at in this enquiry.'

Thus begins a report from academia's expert on all things that grate and are small. John W. Trinkaus, a professor emeritus at New York City's Zichlin School of Business, has turned his gimlet eye to the annoying little aspects of modern life.

Trinkaus won the 2003 Ig Nobel Prize in literature for publishing more than eighty studies of things that annoy him. A former engineer and an ever-curious student of human behaviour, he has personally gathered statistics about people who wear their baseball caps backwards, attitudes towards Brussels sprouts, the marital status of television quiz show contestants, pedestrians who wear sports shoes that are white rather than some other colour, swimmers who swim laps in the shallow end of the pool rather than the deep end, and shoppers who exceed the number of items permitted in a supermarket's express checkout lane. And many, many other quirks of human behaviour. Today his oeuvre totals more than one hundred monographs.

The Trinkaus method is to observe, and then to produce a no-nonsense report, typically two or three pages long. Many of his publications show a deep interest in waiting, obstruction, and delay, as epitomized in his 1985 single-page 'Waiting Times in Physicians' Offices: An Informal Look'.

Waiting is also the subject in Trinkaus's gift of a study 'Visiting Santa: An Additional Look', which he bestowed on us – all of us – in 2007. That was a sequel to the previous year's 'Visiting Santa: Another Look', which built upon the work he described in the very first of his Santa-related studies, 2004's 'Santa Claus: An Informal Look'.

Each of these reports gives a cheerfully dreary look at the behaviour of children and their parents in a shopping mall. As the 2007 report describes it: "The observer [which is to say, Trinkaus] positioned himself unobtrusively a short distance away from a single line of children and guardians advancing to visit with Santa Claus, in a place where the children's and guardians' facial expressions could be noted.' The findings, he writes, are 'consistent with the conclusions that the greater percentage of children appeared indifferent to their visit to Santa'. As in his previous investigations, many of the guardians did look excited, or at least looked like they were trying to look excited.

Time and Guardian's Face	Children			
	Sad	Indifferent	Happy	Total
Week After Thanksgiving and Week Before Christmas: Facial Frequencies of Guardians and Children by Expression				
Week After Thanksgiving				
Sad	0	0	0	0
Indifferent	2	8	1	11
Happy	21	103	15	139
Total	23 (15%)	111 (74%)	16 (11%)	150 (100%)
Week Before Christmas				
Sad	0	2	0	2
Indifferent	5	8	12	25
Happy	1	79	43	123
Total	6 (4%)	89 (59%)	55 (37%)	150 (100%)

Trinkaus also shows a special fascination with people's adherence to laws, regulations, and customs. His 'Stop Sign Compliance: An Informal Look', published in 1982, examined how many motorists did – and how many did not – come to a full stop at a particular streetcorner. Trinkaus did follow-up studies at that same intersection in 1983 ('Stop Sign Compliance: Another Look'), 1988 ('... A Further Look'), 1993 ('... A Follow-Up Look'), and 1997 '... A Final Look'). In yet another parallel series of studies, Trinkaus looked at drivers' compliance with a traffic stoplight. Together, these document an unseemly, seemingly unstoppable rise in scofflawism.

To do his shopping cart research, Trinkaus lurked, in a professional manner, at a supermarket. He kept a close but, again, unobtrusive watch on people who entered the store. This all took place during the spring, 'on weekdays when the weather was fair, during the hours of 0900 to 1600'. He paid attention only to those shoppers who cleared their carts of litter prior to doing their shopping. He found that '69% dumped the rubbish into another cart, 26% dropped it on the sidewalk, and 5% deposited it in a trash container.'

Trinkaus sees in these numbers a small warning sign to society: 'Many people espouse such things as the virtues of the golden rule and brotherly love, but ... one might well wonder how much is rhetoric and how much is real. For example, how much social awareness is being exhibited by those folks leaving behind rubbish in their cart for others to cope with? Too, how much communal consciousness is being evidenced by those people who, when finding rubbish in a cart, shift the disposing problem to others?' He says: 'Understanding and measuring real-life, everyday situations, such as that recounted here, could possibly help in unfolding a better understanding of the make-up and operation of present day society.'

Trinkaus, John W. (2004). 'Clearing the Supermarket Shopping Cart: An Informal Look.' *Psychological Reports* 94: 1442–43.
— (2007). 'Visiting Santa: An Additional Look.' *Psychological Reports* 101: 779–83.

THE STRATEGIC JESUS

A whole new side of Jesus is cropping up in the field of decision science, as a rising generation of scholars is taking Jesus to their collective, theoretic, strategic bosom. Their leader, by eminence and example, and perhaps by judicious application of strategy, is a highly promoted man of both intellect and action.

'Jesus the Strategic Leader', by Lt. Col. Gregg F. Martin of the US Army War College in Carlisle, Pennsylvania, was published in the year 2000. It is fifty-one pages long. 'This is not a religious study', Colonel Martin writes, 'it is a practical analysis. If one believes that Jesus was simply a man, and not, as Christians believe, part-man and part-God, then the study reveals how one of history's greatest leaders led. If, on the other hand, one believes that Jesus was God in human form, then the study not only shows how a great human being practiced the art of leadership, but also how God chose to lead. In either case, the student or practitioner of leadership cannot go wrong.'

The report includes a drawing of Martin's 'pyramid model' of Jesus the strategic leader. According to this model, Jesus is a pyramid, resting atop and partially intersecting God. God is a pyramid, too, but with a broader base. A third, inverted pyramid is supported atop Jesus's pyramid. This third pyramid begins with what Martin calls the 'Top Three' disciples (Peter, James, and John) and broadens to include the other apostles, then the disciples and, topping everything, the masses.

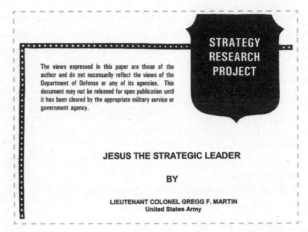

The Strategic Jesus' gives us succinct dictums: 'Develop expertise, then use it with authority ... Choose your battles ... Delegate and power down.'

Colonel Martin left the War College not long after publication of his report. He went on to command the 130th Engineer Brigade of the US Army's Fifth Corps, leading combat engineers before, during, and for more than a year after the invasion of Iraq. He has since returned to his homeland, been promoted to the rank of major general, and been appointed commandant of the War College, where a new generation of American miltary leaders can benefit from his Jesus-style strategic leadership.

Martin, Lt. Col. Gregg F. (2000). 'Jesus the Strategic Leader.' US Army War College strategic report, 5 April, http://handle.dtic.mil/100.2/ADA378218.

BORED MEETINGS

Do you believe, as someone somewhere perhaps does, that meetings, meetings, meetings, followed by more meetings, are altogether a good thing? If so, Alexandra Luong, of the University of Minnesota, Deluth, and Steven G. Rogelberg, of the University of North Carolina at Charlotte, think you should think again. They say: 'We propose that despite the fact that meetings may help achieve work-related goals, having too many meetings and spending too much time in meetings per day may have negative effects on the individual.'

Their report, published in the journal *Group Dynamics: Theory, Research, and Practice,* begins with a somewhat brief recitation of the history of important research discoveries about meetings. Here is a capsule version of their tale.

DISCOVERY: The majority of a manager's typical workday is spent in meetings. This was reported by an investigator named Mintzberg in 1973.

DISCOVERY: The frequency and length of meetings have grown considerably. So declared the team of Mosvick and Nelson in 1987.

DISCOVERY: A scientist named Zohar, in a series of reports published during the 1990s, found evidence that 'annoying episodes' – which are sometimes also known as 'hassles' – contribute to burnout, anxiety, depression, and other negative emotions. Zohar advanced a theoretical framework that may one day help explain why this is so.

DISCOVERY: In 1999, a scientist named Zijlstra 'had a sample of office workers work in a simulated office for a period of two days in order to examine the psychological effects of interruptions. [They] were periodically interrupted by telephone calls from the researcher.' This had what Zijlstra calls 'negative effects' on their mood.

Luong and Rogelberg used those and other discoveries as a basis for their own innovatively broad theory.

They devised a pair of hypotheses, educatedly guessing that:

1) The more meetings one has to attend, the greater the negative effects; and
2) The more time one spends in meetings, the greater the negative effects.

Then they performed an experiment to test their two hypotheses. Thirty-seven volunteers each kept a diary for five working days, answering survey questions after every meeting they attended and also at the end of each day. That was the experiment.

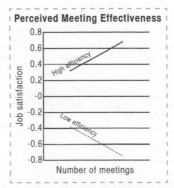

Figure: 'Two-way interaction of number of meetings and perceived
meeting effectiveness to predict job satisfaction'

The results speak volumes. 'It is impressive', Luong and Rogelberg write in their summary, 'that a general relationship between
meeting load and the employee's level of fatigue and subjective
workload was found'. Their central insight, they say, is the concept
of 'the meeting as one more type of hassle or interruption that can
occur for individuals'.

Dr Rogelberg delivered this insight in a talk called 'Meetings and
More Meetings', which he presented to a meeting at the University of
Sheffield. He also does a talk called 'Not Another Meeting!', which
was well received at two meetings in North Carolina and also at two
meetings in Israel.

Luong, Alexandra, and Steven G. Rogelberg (2005). 'Meetings and More Meetings: The Re-
lationship Between Meeting Load and the Daily Well-Being of Employees.' *Group Dynamics:
Theory, Research, and Practice* 9 (1): 58–67.
Rogelberg, S. G., D. J. Leach, P. B. Warr, and J. L. Burnfield (2006). '"Not Another Meeting!"
Are Meeting Time Demands Related to Employee Well-being?' *Journal of Applied Psychology*
91 (1): 83–96.

CORPORATE TIERS OF A CLOWN

Ronald McDonald is not just a clown who hawks hamburgers and
fries. According to two scholars writing in the journal *Leadership
Quarterly*, Ronald McDonald is also a transformational corporate leader.

David M. Boje holds the Bank of America Endowed Professorship
of Management at New Mexico State University. Carl Rhodes is an

associate professor in the School of Management at the University of Technology in Sydney, Australia. Together they produced 'The Leadership of Ronald McDonald: Double Narration and Stylistic Lines of Transformation'.

Boje and Rhodes put their case forthrightly. 'The argument', they say, 'is that rather than just being a spokesperson or marketing device for the McDonald's corporation, Ronald performs an important transformational leadership function.' 'We argue', they argue, 'that while Ronald is crafted by the actual leaders of McDonald's, his leadership exceeds official corporate narratives because of the cultural meanings associated with his character as a clown.'

Clown figures employed by other companies are at best mere employees, at worst mere fictions. As measured by org charts, Mr McDonald towers above the other corporate clowns. Boje and Rhodes reveal that 'since 2003, he has held the quasi-formal executive position of Chief Happiness Officer, and, on 16 April 2004, he became the Ambassador for an Active Lifestyle'.

Boje and Rhodes tell in detail how and why Mr McDonald entered the executive ranks. They then boil it all down to this: 'McDonald's corporate executives believed Ronald could do more than just be a figurehead "spokesclown" at "high-profile public relations stunts such as delivering Happy Meals to the United Nations." [The Russian philosopher Mikhail] Bakhtin's words apply to Ronald: "there always remains in him unrealized potential and unrealized demands".'

Though the researchers may be too modest to suggest it, their Ronald McDonald analysis can be applied to other fields of inquiry. For example, it could help explain recent leadership trends in great nations.

Here are some McNuggets from the study:

Our analysis suggests that a new category of leader is needed; something called a 'clown leader'. As Ronald takes on the ancient masks of rogue, clown, and fool, he integrates diverse forms of laughter (rogue destructive humor, clown merry deception, and fool's right not to comprehend the system). It is this appropriation of a clown type by the world's largest restaurant corporation that is central to its transformation. A method used to transform clowns into leaders is to represent them in adventures of misfortune which are overcome by their leadership powers ...

There is much reason to be skeptical about new forms of leadership that might enhance corporate power in a way that creates new forms of authoritarianism whose operations are far from transparent. This is even more salient for leadership such as Ronald's, whose influence might not easily be noticed given his fictional character.

Boje, David M., and Carl Rhodes (2006). 'The Leadership of Ronald McDonald: Double Narration and Stylistic Lines of Transformation.' *Leadership Quarterly* 17 (1): 94–103.

IN BRIEF

'VISION OF INTEGRATED HAPPINESS ACCOUNTING SYSTEM IN CHINA'
by G. Cheng, Z. Xuand, and J. Xu (published by *Acta Geographica Sinica*, 2005)

A CALCULUS OF PROSTITUTION

There are many theories about prostitution. The theory devised by Marina Della Giusta, Maria Laura Di Tommaso, and Steinar Strøm is one of the few that involves partial differential equations. Sure, they could, if they wished, describe prostitution in words. But for scholars who want to explain prostitution, differential calculus may be the clearest language.

These three scholars, economists all, are based in the UK, Italy, and Norway, making this an international affair. Della Giusta teaches at the University of Reading, Di Tommaso at the University of Turin, and Strøm jointly at the University of Turin and the University of Oslo.

They make their case in a report published in 2007 in the *Journal of Population Economics*. It begins by summarizing the ways in which other economists explain prostitution. Here's a summary of their summary: those other economists are wrong.

The other economists concentrate on gender, pay, and the 'nature of forgone earning opportunities of prostitutes and clients'. But, say Della Giusta, Di Tommaso, and Strøm, those don't count for much. What really matters is the economic role played by stigma and reputation. And the simplest, best way to explain that is with mathematics.

The concepts are as easy and familiar as a good street-corner prostitute:

> U is your satisfaction. It's what you, as a prostitute, really care about – the satisfaction you gain from selling your services. Economists like to call it 'utility', which is why they like to use the letter 'U'.
> L is the amount of leisure you have.
> C is the amount of goods and services you, as a consumer, consume.
> S is the amount of prostitution you, as a prostitute, sell to your customers.
> w is the going price for prostitutes.
> r is a measure of your reputation.

The whole situation, seemingly so complicated, simmers down to a nice partial differential equation. Here it is – Della Giusta, Di Tommaso, and Strøm's rule of thumb for prostitutes. You, a prostitute, find it worthwhile to sell your prostitution services when:

$$[(\partial U/\partial L) / (\partial U/\partial C) \mid S_{p=0}] \le w - [(\partial U/\partial r) / (\partial U/\partial C) \mid S = 0]$$

That's the poetic, simple way of putting it. But prostitution is by tradition considered vulgar, so the team also gives a vulgar, all-words description: 'An individual will start to sell prostitution if the price for selling the first amount of prostitution, minus the costs of a worsened reputation for doing so, exceeds the shadow price of leisure evaluated at zero prostitution sold.'

That, in theory, is the story of prostitution. But it's not the only one.

As with competition among prostitutes, competiton among economists who have theories of prostitution can be spirited. Lena Edlund, of Columbia University in New York, and Evelyn Korn, of Eberhard-Karls-Universität Tübingen in Germany, have also worked up a theory that uses partial differential equations. They call it, modestly, 'A Theory

of Prostitution'. Della Giusta, Di Tommaso, and Strøm cite Edlund
and Korn's theory – but say there are other theories that are 'more
sophisticated'.

Prostitution is a difficult, dangerous profession. In addition to
all of their obvious hardships, prostitutes must also endure the cold,
small knowledge that economists still argue as to exactly what, why,
and how they do their work.

Della Giusta, Marina, Maria Laura Di Tommaso, and Steinar Strøm (2007). 'Who's Watching?
The Market for Prostitution Services.' *Journal of Population Economics* 22 (2): 501–16.
Edlund, Lena, and Evelyn Korn (2002). 'A Theory of Prostitution.' *Journal of Political Economy*
110: 181–214.

SEVEN

OF US, IN BED

IN BRIEF

'PENILE SUFFICIENCY: AN OPERATIONAL DEFINITION'
by C. M. Earles, A. Morales, and W. L. Marshall (published in the
Journal of Urology, 1988)

*Some of what's in this chapter: Quills and copulation • Tipping points of
lap dancing • East Germans hooking up with West Germans • Scholarly
perversion in the twenty first century • The Kiss-Throwing Doll • Beetle
hearts stubbie • Curiousity about castration • Stick figure sex • Gueguen,
breast man • Brindley in plain view • Come now, rhinoceros • You never
sleep alone*

ALBERT AND HIS PORCUPINE PRICKS

How do porcupines make love? Wendy Cooper discovered the answer
while poking around the basement of the Australian National University library in Canberra one day near the turn of the millennium.
Cooper is a parasitologist. She studies parasites, not porcupines. She
also, in the course of her work, studies scientific journals. It was via
this latter method that she acquired her professional knowledge of
porcupines' prickly procreation procedures.

Cooper found two studies written (one with co-authors) by Albert
R. Shadle of the University of Buffalo, New York, dating from 1946.
Shadle was chairman of Buffalo's biology department from 1919 to
1953. One paper is called 'The Sex Reactions of Porcupines (*Erethizon
d. dorsatum*) Before and After Copulation'. The other is 'Copulation in
the Porcupine'.

RATE OF PENETRATION OF A PORCUPINE SPINE

By Albert R. Shadle and Donald Po-Chedley

During the examination of one of the porcupines in the vivarium of the University of Buffalo, the animal became excited, and, as a result of its struggle, fell backwards from the porcelain top of the table. She fell upon the antero-lateral surface of the junior author's right leg, embedding her spines primarily in the belly of the tibialis anterior muscle. Apparently the mid-sacral region of the animal's back made contact with the leg, for the embedded quills were similar in length, diameter, and color to the spines in that area. The force of the fall of the 12.5-pound animal drove its quills through the heavy laboratory coat, the trousers, and deeply into the flesh of the leg. This incident occurred at about 10:50 A.M., January 20, 1947.

Wendy Cooper digested the information and published a summary, carefully worded to make sense both to porcupine specialists and to laypersons. From the outset, she is direct: 'How do porcupines make love? You would probably think the answer is "very carefully", but you would probably be wrong.'

The porcupines in the study were part of a colony that Shadle kept at the University of Buffalo. The colony consisted of five females (Maudie, Nightie, Prickles, Snooks, and Skeezix) and three males (Old Dad, Pinkie, and Johnnie).

Come mating season, the scientists would place a male into a cage that already contained a female. Wendy Cooper describes the subsequent action. First came courtship: 'When the male encountered the female porcupine he smelled her all over, then reared up on his hind legs ... If she was prepared for mating she also reared up and faced the male, belly-to-belly. In this position most males then sprayed the female with a strong stream of urine, soaking her from head to foot. She would either 1) object vocally, 2) strike with her front paws, as though boxing, 3) threaten or try to bite, or 4) shake off the urine and run away. If ready for mating the female did not object strongly to this shower.'

Then the porcupines did their business: 'The male made sexual contact from behind the female. The spines of both animals were relaxed and lay flat. His thrusts were of the 'usual nature' and were produced by flexing and straightening the knees. Males did not grasp the female in any way. Mating continued until the male was exhausted ... If males refused to co-operate, the female approached a nearby male and acted out the male role in coition with the un-involved male.'

This research project was potentially hazardous for the porcu-pines – and for the scientists. But other reports written by Shadle offer perspective.

Shadle's interest in porcupine-human interaction had been long-standing. It began at or about 10:50 a.m. on 20 January 1947: 'During the examination of one of the porcupines in the vivarium of the University of Buffalo, the animal became excited, and, as a result of its struggle, fell backwards from the porcelain top of the table. She fell upon the antero-lateral surface of the junior author's right leg, embedding her spines primarily in the belly of the tibialis anterior muscle. Apparently the mid-sacral region of the animal's back made contact with the leg, for the embedded quills were similar in length, diameter, and color to the spines in that area. The force of the fall of the 12.5-pound animal drove its quills through the heavy laboratory coat, the trousers, and deeply into the flesh of the leg.'

Good scientists do not fail to get good counts and measure-ments. Shadle and his colleague Donald Po-Chedley determined that '79 quills had penetrated the skin deeply enough to secure firm anchorage', and that the deepest penetration was sixteen milli-metres into the leg.

A 1955 paper written by Shadle sums up his two decades of experience. 'Many hundreds of quills have penetrated various parts of the author's own body in numbers of one or two, to as many as forty at one time', he says. 'Usually the fingers, hands or arms were the areas quilled, but on one occasion forty were driven into the forehead and bridge of the nose by one stroke of a porcupine's quill-studded tail, but glasses prevented any injuries to the eyes.' Then, he shares his main discovery: that removing a quill 'is very painful unless done with a quick movement which jerks the quill straight back in the opposite direction from which it entered the flesh'.

'The penetration of porcupine quills into the human body is never a pleasant sensation', Shadle wrote. 'But twenty years of experience in working with a porcupine colony, and continued handling of these spiny animals, have convinced the author that description of the discomfort of being quilled is often very much exaggerated.'

Shadle, Albert R., Marilyn Smelzer, and Margery Metz (1946). 'The Sex Reactions of Porcupines (*Erethizon d. dorsatum*) Before and After Copulation.' *Journal of Mammalogy* 27 (2): 116–21.
Shadle, Albert R. (1946). 'Copulation in the Porcupine.' *Journal of Wildlife Management* 10 (2): 159–62.
— (1955). 'Effects of Porcupine Quills in Humans.' *American Naturalist* 89 (844): 47–49.
Shadle, Albert R., and Donald Po-Chedley (1949). 'Rate of Penetration of a Porcupine Spine.' *Journal of Mammalogy* 30 (2): 172–73.

MAY WE RECOMMEND

'OVULATORY CYCLE EFFECTS ON TIP EARNINGS BY LAP DANCERS: ECONOMIC EVIDENCE FOR HUMAN ESTRUS?'

by Geoffrey Miller, Joshua M. Tybur, and Brent Jordan (published in *Evolution and Human Behavior*, 2007, and honoured with the 2008 Ig Nobel Prize in economics)

The authors, at the University of New Mexico, explain: 'All women made less money during their menstrual periods, whether they were on the pill or not. However, the normally cycling women made much more money during estrus (about US $354 per shift) – about US $90 more than during the luteal phase and about US $170 more than during the menstrual phase. Estrous women made about US $70 per hour, luteal women made about US $50 per hour, and menstruating women made about US $35 per hour. By contrast, the pill users had no mid-cycle peak in tip earnings ... This also results in pill users making only US $193 per shift compared to normally cycling women making US $276 per shift – a loss of more than US $80 per shift.'

GERMAN SEXUAL UNIFICATION

A study called 'The Sexual Unification of Germany' tells what happened, on paper and in some people's heads, when East Germans hooked up with West Germans.

After the Berlin Wall came tumbling down in 1989, salacious minds wondered how many, how quickly, how often, and just plain how Easterners would fall into bed with Westerners.

Ingrid Sharp, a senior lecturer in German at the University of

Leeds, pored through newspapers and academic papers in search of something related to the answer. She published her findings in a 2004 issue of the *Journal of the History of Sexuality*.

Sharp focused on a single question: 'What happened to GDR [German Democratic Republic] sexuality when it was confronted with the sexual mores of West Germany?' 'The answer', she writes, 'appears to have been an explosion of discourse surrounding sex.' In other words: lots of talk, not much action.

In press accounts though, the joint was jumping. For a little while, anyway. Sharp describes one of the main storylines: 'While the traditional behavior of conquering armies (killing the men and raping the women) was obviously inappropriate for Western men after the collapse of Communism, something very similar seemed to be happening on a metaphorical level ... The context was the ideological battle between East and West, the cold war being slogged out in the arena of sexuality, with orgasmic potential replacing nuclear capacity.'

The tabloid press enjoyed a circulation-boosting 'brief obsession with GDR sexuality'. And a grand yet debilitating obsession it was: 'GDR women were represented as products for the fantasies of Western men, while the East German men were dismissed as both socially and sexually inadequate.' Sharp also recounts a West German man's televised claim that 'GDR women are not really uglier than West German women, and they dress as well. But the real advantage is that they are more [modest, undemanding, easily satisfied]'.

On the other side of the former fence, a GDR sexologist named Dr Kurt Starke 'linked findings about women's greater sexual enjoyment to the social policies of the GDR'. The daily tabloid *BILD Zeitung* countered with the headline 'Do GDR Women Really Come More Often? The Orgasm Professor Is Talking Rubbish' above a story in which an East German nurse named Adelheid said: 'We really don't have more orgasms in the GDR. Not me, anyway, because I have to work up to 12 hours a day and that doesn't leave much time for love.'

For the most part, Sharp tells of two Germanies united first by hype about sex, and then by disappointment as most people's sex lives, no matter where they lived, remained humdrum.

The report ends with a deflating comment from journalist Regine Sylvester, who tried to sum up both her own experience and that of the entire nation. The supposed 'sex boom' that happened right after unification, Sylvester opined, 'did not turn the Federal Republic into a noisily copulating society, nor did the official taboos turn the old GDR into an ascetic one'.

Sharp, Ingrid (2004). 'The Sexual Unification of Germany.' *Journal of the History of Sexuality* 13 (3): 348–65.

A CATALOGUE OF PERVERSE BEHAVIOUR

Perversions get a new lease on life, at least chronologically, whenever a new century begins. William L. Salton, a New York City clinical psychologist, rang out the old and rang in the new by writing a study called 'Perversion in the Twenty-First Century: From the Holocaust to the Karaoke Bar'. It appeared in 2004 in the *Psychoanalytic Review*.

After describing some of the many psychological theories about the differences between perversions and non-perversions, Salton in essence takes a cold shower and shakes his head. '[I will] attempt neither to disprove nor to contradict the theories cited in the preceding sections', he writes. 'Instead, I hope to augment and combine them.'

He attempts this by sharing the story of a patient who reluctantly came under his care: 'The patient, whom I will call "Alan", is a 28-year-old male of Gypsy descent. He was referred by the criminal court following repeated convictions for stealing complimentary bathrobes from the rooms of upscale hotels.'

Alan's lawyer repeatedly 'was able to plea bargain probation and psychological counseling, rather than incarceration, when it was determined that Alan did not take the bathrobes to sell them, or to steal whatever contents a guest might have left inside. Instead, he brought them home in order to masturbate into them. He would then discard the bathrobe when it no longer held his sexual interest, thus requiring him to stalk and steal again.'

Alan also had a goal to perform karaoke in a bar in all fifty American states. In short, Alan had some problems. Having told us about them, Salton augments and combines a variety of traditional psychological theories, trying to devise a treatment.

Salton also celebrates some of his predecessors. 'Perversion', he observes cheerily, 'has always been of interest to mental health professionals.' He writes most admiringly about a study that is, roughly speaking, a much grander, twentieth-century equivalent of the one he is preparing. Richard von Krafft-Ebing's 452-page book *Psychopathia Sexualis*, published in 1906, helped give birth to the modern scholarly approach to perversion. Salton says the book 'fascinated psychotherapists and theoreticians alike', being fundamentally 'a catalogue of perverse behaviors and practices that would rival anything on today's Internet'. The book also introduced new words (most influentially 'sadism' and 'masochism'). It sported a delightful index that could teach a thing or two even to non-German readers. Here are three swatches from that index:

> *Dementia paralytica*
> *Diebstahl auf Grund von Fetischismus*
> *Effeminatio*
>
> *Kohabitation*
> *Koketterie*
> *Konträre Sexualempfindung*
>
> *Melancholie*
> *Menstruation*
> *Metamorphosis sexualis paranoica*
> *Misshandlung von Weibern*

Despite these bows to the past, Salton's twenty-first-century study is primarily about poor Alan the complimentary bathrobe thief. Much of Alan's inner world, Salton writes, 'remains a mystery ... I hope and look forward to having a chance to write about Alan's further development in treatment. I plan to call the next article: "From the Karaoke Bar to the Depressive Position".'

That follow-up study has yet to make its appearance before an eager public.

Salton, W. L. (2004). 'Perversion in the Twenty-First Century: From the Holocaust to the Karaoke Bar.' *Psychoanalytic Review* 91 (1): 99–111.

AN IMPROBABLE INNOVATION

'KISS-THROWING DOLL'
by William B. Nutting (US Patent no. 3,603,029, granted 1971)

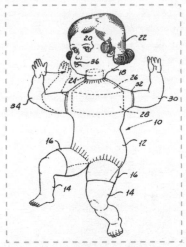

The patented Kiss-Throwing Doll

Anyone who tries to patent a kiss (the romance book publisher Har-lequin, for one, which in February 2011 filed an application for 'The Essential Romantic Kiss') must contend with Nutting. His invention features (says Mr Nutting) 'a central shaft upon which I mount a rotatable driving spool for oscillatory movement under the influence of a drawstring in one direction and under the bias of a coils spring in the other direction'. Metaphorically, has anyone every devised a more perfect description of the romantic kiss?

BEETLE BOTTLE SHAG

Certain Australian males are physically attracted to one particular type of beer bottle. An experiment in Western Australia demonstrated that beer bottles known there as 'stubbies' get recycled in an un-anticipated way. Stubbies are squat little bottles, 370 millilitres in capacity. A study published in 1983 begins with the statement: 'Male

Julodimorpha bakewelli White were observed attempting to copulate with beer bottles.'

Julodimorpha bakewelli (white) are beetles. Prior to 1983, few people were aware that the beetles were having their way with the stubbies. It is still not common knowledge.

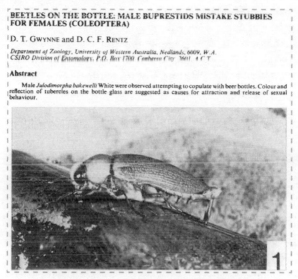

BEETLES ON THE BOTTLE: MALE BUPRESTIDS MISTAKE STUBBIES FOR FEMALES (COLEOPTERA)

D. T. GWYNNE and D. C. F. RENTZ

Department of Zoology, University of Western Australia, Nedlands, 6009, W.A.
CSIRO Division of Entomology, P.O. Box 1700 Canberra City 2601 A.C.T.

Abstract

Male *Julodimorpha bakewelli* White were observed attempting to copulate with beer bottles. Colour and reflection of tubercles on the bottle glass are suggested as causes for attraction and release of sexual behaviour.

Darryl Gwynne, then of the University of Western Australia (he has since moved to the University of Toronto), and David Rentz, of the Commonwealth Scientific and Industrial Research Organisation (CSIRO) in Canberra, tried to alert the world. They published two reports filled with intense but delightful technical detail. 'On two occasions a flying male was observed to descend to a stubbie and attempt to copulate', they write. 'A search was made for other stubbies in the area and two others, with associated beetles, were located. The males were either at the side or "mounted" on top of the bottle, with genitalia everted and attempting to insert the aedeagus. Only one stubbie without a beetle was located. A short experiment was conducted in which four stubbies were placed on the ground in an open area.'

The experiment was a success. The beetles loved the bottles. Gwynne and Rentz later tried to prise them off, but found this not easy to do. One beetle, they observed, was so attached to its bottle that it stayed faithful despite being attacked and gnawed by ants. Gwynne and Rentz witnessed two deaths.

The scientists developed a theory that explains the nature of the seemingly unnatural attraction: 'It was apparent that it wasn't any remaining contents in the stubbies that attracted the beetles; not only do Western Australians never dispose of a bottle with beer still in it but many of the bottles had sand and detritus accumulating over many months ... The brown glass of the stubbies bore a resemblance to the coloration of the beetle; in addition, the rows of regularly-space tubercules on the top and bottom of the bottle reflected the light in a similar way to the pits on the elytra of the beetle.'

Gwynne and Rentz issued a warning to their fellow citizens: 'Improperly disposed of beer bottles not only present a physical and "visual" hazard in the environment, but also could potentially cause great interference with the mating system of a beetle species.'

They say this beetle behaviour is consistent with other biological reports – that in most species it is the male that makes 'mating mistakes'. Their second report, published a year after the first, mentions that a nurse in Perth told them 'a parallel tale involving a male *Homo sapiens* who had entered hospital "attached" to a milk bottle'.

Gwynne, D. T., and D. C. F. Rentz (1983). 'Beetles on the Bottle: Male Buprestids Mistake Stubbies for Females (*Coleoptera*).' *Journal of the Australian Entomological Society* 22: 79–80.
— (1984). 'Beetles on the Bottle.' *Antenna: Proceedings (A) of the Royal Entomological Society London* 8 (3): 116–17.

WHAT COMES AFTER A FROG IN THE THROAT

As a young man, Richard Wassersug talked eleven other scientists into taste-testing tadpoles. His 1971 paper called 'On the Comparative Palatability of Some Dry-Season Tadpoles from Costa Rica' explained why he did it: to answer an old scientific mystery, namely, why don't flashy-looking animals all get eaten by predators and thus go extinct? The answer, it turns out, is that many gaudy tadpoles taste revoltingly bad.

Wassersug later became a professor of biology at Dalhousie University in Halifax, Nova Scotia, and a recognized authority on amphibian physiology and medicine. But, as happens to a few great scientists, a most unexpected, most unhappy event gave his life – and his research – a spectacular turn.

In 1998, at age fifty two, Wassersug was diagnosed with prostate cancer. As a scientist he understood that the medical therapies available for treatment technically made him a eunuch. He discovered that being a eunuch has, in addition to the famous drawbacks, some unexpected benefits: less tendency to be aggressive and pugnacious; more and maybe deeper empathy with other people; and, without the hormone-driven sexual distraction, a relaxed comfort in savouring the beauty in women's faces.

Science mostly overlooks eunuchs as a subject of interest and as a likely source of valuable insights. Wassersug set out to correct that, turning his apparent tragedy into an ecstatic new obsession. His eunuch studies, many done in collaboration with other scientists, include appraisals of medico-social dilemmas. There are also some wonder-filled looks at unfamiliar parts of the human condition.

Wassersug's research took him down a side path where academics seldom tread. There are men who, for reasons unrelated to illness or injury, want to be castrated. Wassersug sought to understand this scientific mystery. In 'A Passion for Castration: Characterizing Men Who Are Fascinated with Castration, but Have Not Been Castrated', which was published in the *Journal of Sexual Medicine,* he and his colleagues 'identify factors that distinguish those who merely fantasize about being castrated from those who are at greatest risk'. A separate study, 'Eunuchs in Contemporary Society: Expectations, Consequences and Adjustments to Castration', reports what may be the most disturbing fact about the voluntary castrati: 'The majority of castrations (53%) were not performed by medical professionals.'

For his tadpole tasting, Wassersug was awarded an Ig Nobel Prize in the field of biology in the year 2000.

Wassersug, Richard (1971). 'On the Comparative Palatability of Some Dry-Season Tadpoles from Costa Rica.' *American Midland Naturalist* 86 (1): 101–9.

Roberts, Lesley F., Michelle A. Brett, Thomas W. Johnson, and Richard J. Wassersug (2007). 'A Passion for Castration: Characterizing Men Who Are Fascinated with Castration, but Have Not Been Castrated.' *Journal of Sexual Medicine* 5 (7): 1669–80.

Brett, Michelle A., Lesley F. Roberts, Thomas W. Johnson, and Richrard J. Wassersug (2007). 'Eunuchs in Contemporary Society: Expectations, Consequences and Adjustments to Castration. Part II' *Journal of Sexual Medicine* 4 (4): 946–55.

SALACIOUS OVER STICK FIGURES

Which body parts do students pay attention to when they size up their rivals in romance? Pieternel Dijkstra and Bram Buunk went to a university library to find the answer. They handed out survey forms to students who were there studying books or studying each others' body parts. The monograph 'Sex Differences in the Jealousy-Evoking Nature of a Rival's Body Build', published in 2001 in the journal *Evolution and Human Behavior*, explains what Dijkstra and Bruunk learned from this endeavour.

The two psychologists, based at the University of Groningen in the Netherlands, begin by reviewing the state of knowledge in their field. Everyone's ultimate goal: clear up the mysteries of romantic rivalry and jealousy.

Previous researchers, Dijkstra and Bruunk say, established that jealousy is 'elicited when people perceive threats to relationships with their partners due to actual or imagined rivals'. Studies suggested that 'people tend to compare rivals' qualities with their own'.

Intending to build on those discoveries, Dijkstra and Buunk made inquiries of some ninety-one women and ninety-four men. They presented their subjects with a survey form that included some stick figure drawings and some written questions. Women were instructed to look at stick figures representing women, men at stick figures of men. The drawings, all with 'identical facial and bodily features', showed a variety of big and small shoulders, waists, and hips. The students were asked to look at each drawing and say 'how jealous they would feel if that individual were romantically interested in their partner'. Students then had to 'rate how attractive' they found each stick figure to be, and how attractive it would be to their actual or imagined romantic partner.

Finally came another question. The study says, dryly: 'When participants had answered all of the above items, they were asked to list the characteristics of the figures' body they had focused upon when answering the questions.' Women mentioned looking at rival women's waist, hips, and legs. Men mentioned looking at rival men's shoulders, chest, and bellies. Women indicated that small-waisted, big-hipped rival female stick figures are 'more socially dominant and attractive'. Men, though, said they were struck by the attractiveness and social dominance of big-shouldered male line drawing rivals, especially those with small waists.

That 2001 paper established the basic facts about how a group of people in their early twenties say they react to stick figure representations of romantic rivals.

But Dijkstra and Buunk kept at it, researching and publishing extensively on assorted aspects of the jealousy / narrow waists / broad shoulders / big hips relationship. Their 2005 monograph, 'A Narrow Waist Versus Broad Shoulders: Sex and Age Differences in the Jealousy-Evoking Characteristics of a Rival's Body Build', tells how this played out with older folks. Like any carefully prepared scientific paper, it forthrightly declares that it has limitations. These are highlighted by the statement: 'in real life individuals are usually not confronted with rivals so scarcely dressed as the figures in our manipulation'.

Dijkstra, Pieternel, and Bram P. Buunk (2001). 'Sex Differences in the Jealousy-Evoking Nature of a Rival's Body Build.' *Evolution and Human Behavior* 22 (5): 335–41.

Buunk, Bram P., and Pieternel Dijkstra (2005). 'A Narrow Waist Versus Broad Shoulders: Sex and Age Differences in the Jealousy-Evoking Characteristics of a Rival's Body Build.' *Personality and Individual Differences* 39 (2): 379–89.

PROFESSOR GUEGUEN, BREAST EFFECTS ANALYST

Professor Nicolas Gueguen finds significance, or at least fascination, in what might be called voyeuristic microscopy, watching how people react to mundanely noticeable sights and sounds and touching. Many of his experiments involve young female confederates who are shaped or perfumed or who lay a hand upon strangers in particular ways. Generally, the test subjects who respond most vigorously are men.

Based at the University of Bretagne-Sud, in Brittany, France, Gueguen has been pumping out publications since the year 2000. He honours the academic custom of referring to himself in print with the royal 'we'.

His experiments probe a range of human behaviour. A study called 'Women's Bust Size and Men's Courtship Solicitation', published in the journal *Body Image*, describes how Gueguen tested 'the effect of a woman's breast size on approaches made by males. We hypothesized that an increase in breast size would be associated with an increase in approaches by men.' The study ends with an 827-word ode on the topic sentence: 'Our hypothesis was confirmed.'

A related experiment produced a study called 'Bust Size and Hitchhiking: A Field Study', published in *Perceptual and Motor Skills*. There Gueguen reports that '1200 male and female French motorists were tested in a hitchhiking situation. A 20-year-old female confederate wore a bra which permitted variation in the size of cup to vary her breast size. She stood by the side of a road frequented by hitchhikers and held out her thumb to catch a ride. Increasing the bra-size of the female-hitchhiker was significantly associated with an increase in number of male drivers, but not female drivers, who stopped to offer a ride.'

Frequency and Percent of Motorists Who Stopped by Experimental Condition and Sex of Motorist

Motorists' Sex	n	Bra Cup Size		
		A	B	C
Male	774	14.92%	17.79%	24.00
		40/268	46/256	60/250
Female	426	9.09%	7.64%	9.33%
		12/132	11/144	14/150

An earlier study called 'The Effect of Touch on Tipping: An Evaluation in a French Bar', published in the *International Journal of Hospitality Management*, aimed to fill a very specific gap in psychologists' knowledge of human behaviour. The study explains: 'Although positive effect of touch on restaurant's tipping has been widely found in the literature, no evaluation was made outside the United States of America and in a bar. An experiment was carried out in a French bar. A waitress briefly touched (or not) the forearm of a patron when asking him/her what he/she wanted to drink. Results show that touch increases tipping behavior although giving a tip to a waitress in a bar is unusual in France.' Gueguen has pursued related questions, some involving smiles, upon which he reports in additional studies.

'The Effect of Perfume on Prosocial Behavior of Pedestrians', published in the journal *Psychological Reports*, is representative of several Gueguen investigations of how people respond to the presence and actions of a heavily perfumed woman. In this one, the fragranced woman walks in front of strangers and 'drops a packet of paper handkerchiefs or a glove apparently without noticing'.

Into these and other forays does Professor Gueguen probe and ponder the human condition.

Gueguen, Nicolas (2007). 'Women's Bust Size and Men's Courtship Solicitation.' *Body Image* 4 (4): 386–90.

— (2007). 'Bust Size and Hitchhiking: A Field Study.' *Perceptual and Motor Skills* 105 (3): 1294–98.

— (2001). 'The Effect of Perfume on Prosocial Behavior of Pedestrians.' *Psychological Reports* 88: 1046–48

—, and Celine Jacob (2005). 'The Effect of Touch on Tipping: An Evaluation in a French Bar.' *International Journal of Hospitality Management* 24 (2): 295–99.

IN BRIEF

'SUCCESSFUL INSEMINATION EXPERIMENTS WITH CRYO-PRESERVED SPERM FROM WILD BOARS'

by D. Krause, D. Ick, and H. Treu (published in *Zuchthygiene*, 1981)

A STIFF TEST

Giles Skey Brindley, MD, FRCP, FRS, knows how to stand proud. At a 1983 Urodynamics Society lecture in Las Vegas, Dr Brindley demonstrated – with panache – that he could inject drugs into his penis and thereby cause an erection, and a stir.

Brindley had developed the first effective treatment for what was then loosely called 'impotence' and today goes by the stiffer euphemism 'erectile dysfunction'. His appearance in Las Vegas ensured that the discovery would not go unnoticed.

Two decades later, Laurence Klotz, a University of Toronto urologist, wrote a firsthand account of his experience at that meeting. Entitled 'How (Not) to Communicate New Scientific Information: A Memoir of the Famous Brindley Lecture', it enlivens and to some extent graces the pages of the urological journal *BJU International*. Klotz reports that Brindley 'indicated that, in his view, no normal person would find the experience of giving a lecture to a large audience to be erotically stimulating or erection-inducing. He had, he said, therefore injected himself with papaverine in his hotel room before coming to give the lecture, and deliberately wore loose clothes to make it possible to exhibit the results ... He then summarily dropped his trousers and shorts, revealing a long, thin, clearly erect penis. There was not a sound in the room. Everyone had stopped breathing. But the mere public showing of his erection from the podium was not sufficient. He paused, and seemed to ponder his next move. The sense of drama in the room was palpable. He then said, with gravity, "I'd like to give some of the audience the opportunity to confirm the degree of tumescence". With his pants at his knees, he waddled down the stairs, approaching (to their horror) the urologists and their partners in the front row.' And so on.

Brindley's activities range wide in science and medicine, and also in music. He invented a new variety of bassoon, and in 1973 brought to bear many of his diverse interests in a treatise in the journal *Nature* called 'Speed of Sound in Bent Tubes and the Design of Wind Instruments'.

The self-injection erection experiment entered the medical literature in 1986, in the March issue of the *British Journal of Pharmacology*, in the form of Brindley's treatise 'Pilot Experiments on the Action of Drugs Injected into the Human Corpus Cavernosum Penis'. Brindley writes: 'Drugs were injected through a 0.5 millimeter x 16 millimeter needle into the right corpus cavernosum in the proximal third of the free penis. The penis was then massaged systematically to distribute the drug throughout both corpora cavernosa as follows...'. There follows a 307-word description of the drugs and of the massage technique.

The final word can be left to Klotz, who says: 'Professor Brindley belongs in the pantheon of famous British eccentrics who have made spectacular contributions to science. The story of his lecture deserves a place in the urological history books.'

Klotz, Laurence (2005). 'How (Not) to Communicate New Scientific Information: A Memoir of the Famous Brindley Lecture.' *BJU International* 96 (7): 956–57.
Brindley, Giles S (1973). 'Speed of Sound in Bent Tubes and the Design of Wind Instruments.' *Nature* 246: 479–80.
— (1968). 'The Logical Bassoon.' *Galpin Society Journal* 21: 152–61.
— (1986). 'Pilot Experiments on the Action of Drugs Injected into the Human Corpus Cavernosum Penis.' *British Journal of Pharmacology* 87 (3): 495–500.

DISTRACTEDLY, DECIDEDLY AROUSED

When a young man masturbates, exactly how distracted does he get? An experiment performed on students at the University of California, Berkeley aimed to find out.

Full details were published in the *Journal of Behavioral Decision Making*. Dan Ariely, of the Massachusetts Institute of Technology, and George Loewenstein, of Carnegie Mellon University in Pittsburgh, describe their arousing achievement in dry, formal terms: 'We examine the effect of sexual arousal, induced by self-stimulation, on judgments and hypothetical decisions made by male college students.'

The scientists begin their report by pointing out that 'sexual motivation plays a direct or indirect role in wide-ranging social interactions and in considerable economic activity.' Pornography alone, they say, takes in more revenues in the United States than the three largest professional sports (football, basketball, and baseball) combined.

Having established that the topic is of value, Ariely and Loewenstein get right to the action. They explain how they recruited thirty-five students, offering to pay each a small fee for the effort of masturbating while answering a survey. Each student was given a laptop computer with a keypad 'designed to be operated easily using only the non-dominant hand'.

Some of the volunteers had instructions to answer the questions 'while in their natural, presumably not highly aroused, state'. Others 'were first asked to self-stimulate themselves, and were presented with the same questions only after they had achieved a high but suborgasmic level of arousal'.

The computer screen displayed 'an "arousal thermometer" with regions coloured from blue to red representing increasing levels of arousal. Two keys on the keypad allowed the user to move the probe on the arousal meter to indicate their momentary level of arousal. The panel on the top-left occupied the largest part of the screen, displaying diverse erotic photographs.'

The screen also showed the long series of survey questions. Some questions asked about the attractiveness of different sexual activities, items, and opportunities. Among them: women's shoes; a twelve-year-old girl; an animal; a fifty-year-old woman; a man; and an extremely fat person. Other questions probed the risks the volunteer would take in order to obtain sexual gratification.

The volunteers were instructed to press the computer's tab key if they ejaculated. None reported doing so.

Ariely and Loewenstein say their results are 'striking' and

more than confirm what most people believe about young men as a group – that when aroused, they (1) become sexually attracted to things otherwise offputting; (2) grow more willing to engage in morally questionable behaviour that might lead to sex; and (3) are more likely to have unprotected sex. '[Our] study shows that sexual arousal influences people in profound ways', they write. 'Efforts at self-control that involve raw willpower are likely to be ineffective.'

This is a dig at theorists – the ones who advise people to 'just say no' – from experimentalists who are unafraid to get their hands dirty.

(In 2008, Dan Ariely and three of his colleagues received the Ig Nobel Prize in medicine for their study on the effects of expensive fake medicines versus cheap fake medicines.)

Ariely, Dan, and George Loewenstein (2006). 'The Heat of the Moment: The Effect of Sexual Arousal on Sexual Decision Making.' *Journal of Behavioral Decision Making* 19: 87–98.

HORN OF PLENTY

'Monitoring Electroejaculation in the Rhinoceros with Ultrasonography' is the title of a research study published in 1996. The study is notable – deserves and, perhaps, demands attention – for at least two reasons. First, because of its subject.

The main author, Nan Schaffer, is a Chicago-based veterinarian. She published this report in the same year she founded a non-profit organization called SOS Rhino. The group tries to keep the world's five rhinoceros species from becoming extinct. Schaffer was, and is, one of the foremost researchers in the field of rhinoceros reproduction. The field receives little public acclaim.

In the rhinoceros, reproduction occurs, if at all, through a two-part process. First, a male produces semen. Then the semen is transported into a female. The process often goes awry.

ULTRASONOGRAPHIC MONITORING OF ARTIFICIALLY STIMULATED EJACULATION IN THREE RHINOCEROS SPECIES (*CERATOTHERIUM SIMUM, DICEROS BICORNIS, RHINOCEROS UNICORNUS*)

Nan Schaffer, D.V.M., William Bryant, D.V.M., Dalen Agnew, D.V.M., Tom Meehan, D.V.M., and Bruce Beehler, D.V.M.

Abstract: Manual massage of the penis and rectal electroejaculation methods have been minimally effective for collecting semen from the rhinoceros. These two methods for stimulating ejaculation were evaluated by rectal ultra-

Vets try to lend a helping hand. Sometimes this is literally true, and sometimes it involves the use of electromechanical devices. In a 1998 study, Schaffer and her colleagues explain that 'manual massage of the penis and rectal electroejaculation methods have been minimally effective for collecting semen from the rhinoceros'. That pretty much summed up the state of the art. And that art, of course, applied to just part one of the two-part basic reproductive process.

This is dangerous work, less for the animals than for the humans. Much less. A human typically weighs only one-tenth, and in some cases one-fortieth, as much as a grown male rhinoceros. The difference in heft is abetted by the spirited muscular potential of a male rhinoceros as it is being stimulated, directly or indirectly, by the exertions of rhinoceros reproduction technicians. This is also painstaking work, requiring careful engineering, to be performed always and only with utmost caution. The veterinarians and their assistants who engage in this activity do so with a small array of specialized, carefully developed procedures and equipment. The rhino's mighty horn adds, pointedly, to the peril. (The name rhinoceros is derived from Greek words that translate to 'horned nose', a description that lends itself to punnery in connection with any discussion of rhinoceros reproduction.)

But back to Schaffer's 1996 study.

The title 'Monitoring Electroejaculation in the Rhinoceros with Ultrasonography' grabs your attention. But it is not the most significant thing about the report.

To appreciate the most significant part, think back to any writing course you ever had, at any level of school. Almost certainly the teacher told you, and repeated many times, a basic piece of advice: when writing a report, it is important to have a good lead sentence.

Here is the lead sentence in Dr Schaffer's report. Read it aloud. It says: 'Electroejaculation is difficult to perform in the rhinoceros.'

I recommend that whenever you write a report – no matter what the subject – begin it with that sentence.

Schaffer, Nan, Tom Meehan, William Bryant, and Dalen Agnew (1996). 'Monitoring Electro-ejaculation in the Rhinoceros with Ultrasonography.' *Proceedings of the Annual Meeting of the Society for Theriogenology*, Kansas City, Missouri, August.

Schaffer, N., W. Bryant, D. Agnew, T. Meehan, and B. Beehler (1998). 'Ultrasonographic Monitoring of Artificially Stimulated Ejaculation in Three Rhinoceros Species (*Ceratotherium Simum, Diceros Bicornis, Rhinoceros Unicornus*).' *Journal of Zoo and Wildlife Medicine* 29 (4): 386–93.

THE EFFECT OF MOBILE PHONES ON RABBIT SEX

Lest anyone wonder why four scientists studied the effect of mobile phones on rabbits' sex lives, Nader Salama, Tomoteru Kishimoto, Hiro-Omi Kanayama, and Susumu Kagawa spelled out their reasons. Many scientists had tried (though for the most part failed) to prove that repeatedly holding a mobile phone against a person's head causes damage to the brain. The four scientists looked ahead to a perhaps different question: will holding a mobile phone near a man's testicles affect that man's sexual behaviour?

They devised an experiment. Given the expense, complexity, and delicacy of doing it with humans, they opted instead for rabbits.

Salama et al. say, sweepingly, that they are the first to 'have analyzed the potential effect of exposure to electromagnetic waves emitted from mobile phones on male sexual behavior'. Details appear in their monograph called 'Effects of Exposure to a Mobile Phone on Sexual Behavior in Adult Male Rabbit: An Observational Study', published in the *International Journal of Impotence Research*. The team performed this experiment at Tokushima School of Medicine in Japan.

They documented the ruttings (under admittedly artificial conditions) of six male rabbits that had switched-on phones placed near their genitals for twelve weeks, six that had switched-off phones, and another six that were phoneless.

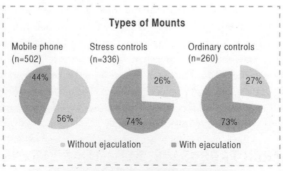

Figure: Types of mounts, with and without mobile phone

The scientists noted the particulars of each mounting, and watched for the moment each rabbit went into 'a state of sexual exhaustion'. They report that the bunnies with active phones 'got sexually exhausted earlier'. This discovery, they emphasize, 'might have some practical implications'. Research in urology and impotence often involves the interplay — sometimes delicate, sometimes not — of technology and biology. The team knows this well. Three years earlier, they published a report with the mostly self-explanatory title Unusual Trivial Trauma May End With Extrusion of a Well-Functioning Penile Prosthesis: A Case Report. It presents not one, but two instructive cases.

The first concerns a fifty-seven-year-old man who 'claimed the prosthesis had been functioning well, giving him and his two wives, as he had a polygamous marriage, an excellent degrees [sic] of satisfaction'. The problem was that 'he also reported having bumped his penis into the suitcase of the preceding passenger while boarding an airplane five days prior to presentation.'

The second patient was a sixty-four-year-old-man who 'described having trapped his penis against a toilet seat while sitting down to defecate four days earlier'. The doctors removed the device from each sufferer. They matter-of-factly report that 'recovery was uneventful in both cases'. The doctors remark: 'These prostheses [which are 13 millimetres in diameter] are somewhat bulky and cannot be satisfactorily crammed into relatively small organs. This crowding quite possibly invites 'easier exposure of patient organs to unexpected trauma'. The big lesson, says the report, is that: 'When the implantation of a malleable penile prosthesis is considered, appropriate sizing should be taken into account.'

We will now see if, prosthesis-less and left to their own devices, these men, like rabbits, breed.

Salama, Nader, Tomoteru Kishimoto, Hiro-Omi Kanayama, and Susumu Kagawa (2010). 'Effects of Exposure to a Mobile Phone on Sexual Behavior in Adult Male Rabbit: An Observational Study.' International Journal of Impotence Research 22: 127–33.
— (2007). 'Unusual Trivial Trauma May End With Extrusion of a Well-Functioning Penile Prosthesis: A Case Report.' Journal of Medical Case Reports 1: 34.

RESEARCH PROPOSALS: SEX WITH A STRANGER

'Gender Differences in Receptivity to Sexual Offers' should be a screamingly famous research report. Yet most people don't know about it. Or maybe they can't believe it exists.

It exists.

Published in 1989 in the *Journal of Psychology and Human Sexuality*, this seventeen-page sizzler tells a simple story. Five women and four men were sent, one at a time, on to a university campus. Each approached strangers of the opposite sex, and said: 'I have been noticing you around campus. I find you to be very attractive.' They then invited the strangers to have sex.

This experiment was performed twice, once in 1978, and again in 1982. The results were the same. As the report describes it: 'The great majority of men were willing to have a sexual liaison with the women who approach them. Not one woman agreed to a sexual liaison.'

The study was conceived and directed by two psychology professors, Elaine Hatfield of the University of Hawaii at Manoa and Russell D. Clark III of Florida State University. It begins with a declaration: 'According to cultural stereotypes, men are eager for sexual intercourse; it is women who set limits on such activity.' It ends with a declamation: 'Regardless of *why* we secured these data, however, the existence of these pronounced gender differences is interesting.'

The paper never does exactly explain *why* they secured the data, but it does supply a list of fifty-nine earlier published studies that they found useful, interesting, or at least worth listing. These include four other sex-related reports by Hatfield and three technical reports from the prestigious US Commission on Obscenity and Pornography.

Fourteen years later, Hatfield and Clark published a study called 'Love in the Afternoon', in which they tried to explain why they had done the experiment and what happened as a result. Here is a nutshell version of their explanation:

> In the spring of 1978, Russ Clark was teaching a small class in experimental social psychology ... Russ dropped a bomb. 'Most women', he said, 'can get any man to do anything they want. Men have it harder. They have to worry about strategy, timing, and "tricks".'
>
> Not surprisingly, the women in the class were incensed. One woman sent a pencil flying in Russ's direction.
>
> In one of Russ's finer moments, he observed: 'We don't have to fight. We don't have to upset one another. It's an

empirical question. Let's design a field experiment to see who's right!'

Journal after journal refused to publish their paper, giving harsh comments of which this one is typical: 'The study itself is too weird, trivial and frivolous to be interesting. Who cares what the result is to such a silly question.'

But Hatfield and Clark were undaunted. As they explain at the end of 'Love in the Afternoon': 'The trivial, uninteresting, and morally suspect research of today often turns out to be the "classic study" of tomorrow.'

Clark, Russell D. III, and Elaine Hatfield (1989). 'Gender Differences in Receptivity to Sexual Offers.' *Journal of Psychology and Human Sexuality* 2 (1): 39–55.
— (2003). 'Love in the Afternoon.' *Psychological Inquiry* 14 (3–4): 227–31.

BED MATES, ALWAYS

Nobody sleeps alone. This has little or nothing to do with morals. It is simply a law of nature, a fact. Census after census finds that, with or without the niceties of formal marriage, dust mites are the great silent majority in every bed.

Professor J. E. M. H. van Bronswijk, of Eindhoven Technical University in the Netherlands, took a good, long, scientific look at who's in bed with what. Van Bronswijk discussed all the dirty details at a meeting of the Benelux Congress of Zoology in 1994. Her study is called 'A Bed Ecosystem'.

A bed is a crowded place. Even without the people, it is full of biomass. Van Bronswijk wrote that this biomass 'consists of domestic mites (mainly of the family Pyroglyphidae) and domestic fungi (mainly the genera *Apserfillus, Penicillium, Wallemia*), with a smaller contribution of insects, spiders and bacteria'. Mostly, it's mites.

This was exciting news. In the decades since van Bronswijk's charming public pillow talk, many other scientists have taken up the practice of bedroom biological voyeurism.

Krzysztof Solarz, of the Silesian Medical Academy in Katowice, Poland, conducted a study of three beds in Sosnowiec, Upper Silesia. This was, Solarz reported in the *Annals of Agricultural and Environmental Medicine* in 1997, the first such investigation ever done in Poland. The city of Sosnowiec had, at that time, a human population of about 250,000. The number of dust mites was anyone's guess.

Solarz counted mite population samples at different times throughout the year. He then compared these with previously published data from beds in the Czech Republic, the Netherlands, Romania, England, Spain, India, Hawaii, and elsewhere.

Dust mites are not everyone's cup of tea, even though they might be in everyone's cup of tea, if the cup is allowed to sit long enough. For some people, dust mites lack interest – sleeping with them is as far as most folks are willing to go.

> SEASONAL DYNAMICS OF HOUSE DUST MITE POPULATONS IN BED/MATTRESS
> DUST FROM TWO DWELLINGS IN SOSNOWIEC (UPPER SILESIA, POLAND):
> AN ATTEMPT TO ASSESS EXPOSURE
>
> Krzysztof Solarz

For the new enthusiast, though, there is plenty to learn, and no end of good things to read. Anyone who enjoys poetry, even a mite, might do well with H. R. Sesay and R. M. Dobson's 1972 'Studies on the Mite Fauna of House Dust in Scotland with Special Reference to that of Beddings'. For the mite-lover who detests poetry, there is J. Z. Young's 1981 prose masterpiece: 'Morphological Adaptation for Precopulatory Guarding in Astigmatic Mites'.

Acarologists – scientists who study ticks and mites – are, like the objects of their study, happy to gather in groups. Acarologists in search of bed partners, inhuman or otherwise, convene each year at the International Congress of Acarology. You can join them, if you wish. According to past conference organizers, 'We look forward to meeting … anyone with a keen interest in mites and/or ticks.'

Professor van Bronswijk received the 2007 Ig Nobel Prize in the field of biology for her census of our bed mates.

Van Bronswijk, J. E. M. H. (1994). 'A Bed Ecosystem.' Lecture Abstracts – 1st Benelux Congress of Zoology, Leuven, 4–5 November.

Solarz, Krzysztof (1997). 'Seasonal Dynamics of House Dust Mite Populations in Bed/Mattress Dust from Two Dwellings in Sosnowiec (Upper Silesia, Poland): An Attempt to Assess Exposure.' *Annals of Agricultural and Environmental Medicine* 4: 253–61.

Sesay, H. R., and R. M. Dobson (1972). 'Studies on the Mite Fauna of House Dust in Scotland with Special Reference to that of Beddings.' *Acarologia* 14: pp. 384–92.

Young, J. Z. (1981). 'Morphological Adaptation for Precopulatory Guarding in Astigmatic Mites (*Acari: Acaridida*).' *International Journal of Acarology* 18: 49–54.

IN BRIEF

'TRAUMATIC LOVE BITES'
by M. Al Fallouji (published in the *British Journal of Surgery*, 1990)

ANON'S SEX LIFE

Has anyone done scientific research about beards? Well, yes. Most of it concerns beards that are attached to scientists. Most of those researchers are men. Most of them are British. Why that should be, I don't know.

In 1970, the journal *Nature* published a letter called 'Effects of Sexual Activity on Beard Growth in Man'. The author's name was suppressed for reasons that may be self-evident. I'll call him here, as *Nature* did there, 'Anon'. Anon's letter said: 'During the past two years I have had to spend periods of several weeks on a remote island in comparative isolation.'

Anon went on to say that he had measured his beard growth 'by collecting and weighing the shavings from the head of a Philips Philishaver razor after a single shave once every 24 hours'. He learned two things. First, that during the day or so before he resumed sexual activity, his beard grew much faster. Second, that within a day or two after resuming the festivities, his beard growth slowed.

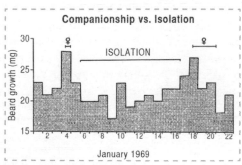

Changes in beard growth during a short stay on the island

Anon's letter provoked an influx of letters to *Nature*, from proud men named Hardisty, Huxley, Bullough, Parsons, Goodhart, and Cook. All were published under the general heading 'Sexual Activity and Beard Growth'. Hardisty, Huxley, Bullough, Parsons, Goodhart, and Cook raised a variety of points, only some of which constitute hair-splitting.

Hardisty inquired whether Anon had been shaving more closely whenever a conjugal meeting was imminent. Huxley questioned whether Anon had consistently measured his beard every day at the same time. Bullough delved into the matter of tension exerted by the follicle erector muscle. Parsons offered a helpful hint about measuring the fluid content of the facial skin. Goodhart vented the opinion that Anon 'ought, in the interests of science, to try abstaining from sexual activity during some of his returns to civilization'. Finally, Cook, who was employed at the Clinic for Nervous Disorders, in London, suggested that 'slight emotional stress may have a stimulating effect on beard growth'.

There have been numerous other scientific reports about beards, many of them fully as consequential as Anon's.

Of perhaps greater import to bearded men is a series of letters that appeared in 1998 and 1999 in the journal *Anaesthesia*. A pair of doctors, Ames and Vincent, at the Queen Victoria Hospital in East Grinstead, in Sussex, England, wrote: 'Maintaining an airway using a facemask in patients who have beards can be difficult ... A simple solution to this problem is to wrap cling film repeatedly around the face and head of the unconscious patient.' This prompted letters from physicians in Sutton Coldfield and Southampton, begging to differ.

Vincent, C., and W. A. Ames (1999). 'The Bearded Airway.' *Anaesthesia* 53 (10): 1034–35.

VORACEK ON THE CENTREFOLD MODEL

Dr Martin Voracek is a shoes-and-ships-and-sealing-wax sort of specialist. His expertise ranges from romance and jealousy, to the 'accuracy of volume measurement in human cadaver kidneys', to the effects of solar eclipses on suicide, and also politics, intelligence, and much else.

A man of many degrees (specifically, DSc, PhD, MSc, and MPh), Voracek is a research resident at the department of psychoanalysis and psychotherapy at the University of Vienna Medical School. His work is known best to, yes, specialists. But he has enjoyed at least two rounds of public notice.

His 2002 study in the *British Medical Journal*, called 'Shapely Centrefolds? Temporal Change in Body Measures: Trend Analysis', is an exercise in statistical voyeurism. Written together with Maryanne Fisher of Canada's York University, it contains language that one might call 'solve-for-x rated'. Here is one passage: 'We looked at the trends in *Playboy* centrefold models' body measurements by analysing 577 consecutive monthly issues, from the magazine's inception in December 1953 to December 2001. We extracted centrefolds' anthropometric data: height, weight, and measurements for bust, waist, and hip. We calculated composite measures from these data: body mass index, waist:hip ratio, waist:bust ratio, bust:hip ratio, and an androgyny index.'

Voracek and Fisher topped themselves four years later in the *Archives of Sexual Behavior* with their study 'Success Is All in the Measures: Androgenousness, Curvaceousness, and Starring Frequencies in Adult Media Actresses'. Its language is sensationally statistical, even in a snippet: 'We retrieved movie and magazine starring frequencies of 125 adult media actresses from a company's database, operationalized starring frequencies as female physical attractiveness measures, and tested their relationship to actresses' anthropometric data.'

Voracek's other writings are, some of them, on duller topics. Here is a little list of representative titles, and of the journals in which they are published:

Digit Ratio (2D:4D), Lateral Preferences, and Performance in Fencing (*Perceptual and Motor Skills*)

Three-Dimensional Histomorphometric Analysis of Distraction Osteogenesis Using an Implanted Device for Mandibular Lengthening in Sheep (*Plastic and Reconstructive Surgery*)

Suicide and General Elections in Austria: Do Preceding Regional Suicide Rate Differentials Foreshadow Subsequent Voting Behavior Swings? (*Journal of Affective Disorder*)

Universal Sex Differences in the Desire for Sexual Variety: Tests from 52 Nations, 6 Continents, and 13 Islands (*Journal of Personality and Social Psychology*)

Patterns and Universals of Mate Poaching Across 53 Nations:
The Effects of Sex, Culture, and Personality on Roman-
tically Attracting Another Person's Partner (*Journal of
Personality and Social Psychology*)
"I Find You to Be Very Attractive...": Biases in Compliance
Estimates to Sexual Offers (*Psicothema*)
Sex and Side Differences in Relative Thumb Length (*Journal
of Hand Surgery*)

There are more – over a hundred so far.

Partik, B. L., A. Stadler, S. Schamp, A. Koller, M. Voracek, G. Heinz, and T. H. Helbich (2002).
'3D versus 2D ultrasound: Accuracy of Volume Measurement in Human Cadaver Kidneys.'
Investigative Radiology 37: 489–495.
Ploder, O., F. Kanz, U. Randl, W. Mayr, M. Voracek, and H. Plenk (2002). 'Three-dimensional
Histomorphometric Analysis of Distraction Osteogenesis Using an Implanted Device for
Mandibular Lengthening in Sheep.' *Plastic and Reconstructive Surgery* 110: 130–37.

MR FOOD SEX

The sexing of food owes much to a gentleman named Ernest Dichter.
As a recent study puts it, 'Dichter was a critical force in encouraging
advertisers to promote a food's sex.' Many psychology and marketing
books call Dichter 'the father of motivation research'.

Today, the Ernest Dichter Institute in Frankfurt, Germany, promotes the name and fame of the master, who spent his professionally
formative years in not-so-distant Vienna, a city that has long fetishized
both sex and food. Dichter prospered, in later years, by advising food
manufacturers to sex up their advertising.

Katherine Parkin, an assistant professor of history at Monmouth
University in New Jersey, published a tribute to Dichterism in the
Advertising and Society Review. She writes that Dichter encouraged
his clients to 'promote foods as feminine or masculine', and that he
'promoted a belief in the sexual qualities of various foods'.

Dichter's pronouncements ranged 'from his insights into Rice
Krispies as a "bubbling, vivacious, young woman" to his advice on
how to masculinize fish'.

He could see the hidden sexual potency in any food. Cake, for
instance. Dichter wrote: 'Perhaps the most typically feminine food is
cake... [Wedding cake is] the symbol of the feminine organ. The act

of cutting the first slice by the bride and bridegroom together clearly stands as a symbol of defloration.' Furthermore, 'women's demand for moistness in a cake reinforced its feminine symbolism'. Women's dislike of dry cake 'may represent a projection onto the cake of the woman's feelings about herself. She wants to be moist and fresh, dewy-eyed and moist-lipped, not a dried up, barren, old crone.'

Dichter recognized the public's appetite for all things Freudian. He let it be known that for twenty years he had lived 'across the street from Sigmund Freud', and that he had taken a public speaking course taught by Freud's daughter-in-law. Dichter took sex to be the key to selling all goods. He counselled car manufacturers that 'a convertible was like a man's mistress and a sedan was like a wife'.

Food, though, seems to have been his special passion. Parkin quotes a 1955 Dichter memo headlined 'Creative Research Memo on the Sex of Rice', which attributes a scientific basis to the sexing of food. It says: 'In an experiment conducted by a famous surgeon, it was discovered that food has sex. While administering barium during the examination of the oesophagus, the good doctor found that, when he mentioned the word "salad" to his female patients, their oesophagi dilated, permitting the passage of the chalky compound. When the word "steak" was suggested to the male patients, their oesophagi reacted similarly.'

In later life, says Parkin, Ernest Dichter focused on food he took to be masculine. Wieners and luncheon meats fascinated him. 'Men', he wrote in 1968, 'do not appear to be as "embarrassed" in eating wieners as women appear to be'.

Parkin, Katherine (2004). 'The Sex of Food and Ernest Dichter: The Illusion of Inevitability' *Advertising and Society Review* 5 (2).

EIGHT

EXCITING INJURIES AND ILLS

IN BRIEF

**'TAKING ACTION ON THE VOLUME-QUALITY RELATIONSHIP: HOW
LONG CAN WE HIDE OUR HEADS IN THE COLOSTOMY BAG?'**
by Thomas J. Smith, Bruce E. Hillner, and Harry D. Bear (published
in the *Journal of the National Cancer Institute*, 2003)

*Some of what's in this chapter: Misnomers of the mouth • The romance of
proctology • Disco dangers • A huge Parisian tooth-yanker • Louis XIV's
missing teeth • Eat your mummy • Michael Jackson surgery • In pursuit of a
wretched itch • Dr Bean's fingernails • Tripping over a black cat • Unlubricated
karaoke • The celebrated rectum of the Bishop of Durham*

A HOT POTATO

Dr Mahmood Bhutta's greatest achievement – measuring the sound
produced by a person with a hot potato in his mouth – has been overlooked in the flurry of attention given his newer study on whether
sexual thoughts can trigger sneezing fits.

Bhutta practises surgery at Wexham Park Hospital, in Slough,
UK. His paper, published in the *Journal of the Royal Society of Medicine*
under the title 'Sneezing Induced by Sexual Ideation or Orgasm: An
Under-Reported Phenomenon', has brought acclaim to Bhutta and
to his co-author and fellow sneezing-induced-by-sexual-ideation-or-
orgasm expert Dr Harold Maxwell, a retired honorary senior lecturer
and consultant psychiatrist formerly at West Middlesex University
Hospital, in Isleworth. The paper brought an elevated level of reportage
to a phenomenon that had appeared in formal medical reports only a

handful of times, in 1875, 1872, and 1972. Trolling through the ailment-infested chat rooms of the Internet, Bhutta and Maxwell unearthed seventeen new cases of people who say they sneeze immediately after thinking about sex, and three others who complain or brag that they sneeze after experiencing an orgasm.

Back in 2006, Bhutta worked in the department of ear, nose, and throat, head and neck surgery at the Royal Sussex county hospital in Brighton. He and fellow Royal Sussex ear-nose-and-throat, head-and-neck surgery specialists George A. Worley and Meredydd L. Harries examined the phenomenon (unrelated to sexual ideation, orgasm, or sneezing) known as 'hot potato voice'.

The study '"Hot Potato Voice" in Peritonsillitis: A Misnomer' appeared in the *Journal of Voice*. 'Voice changes are a well-recognized symptom in patients suffering from peritonsillitis', the authors explain in the report. 'The voice is said to be thick and muffled and is described as a "hot potato voice", because it is believed to resemble the voice of someone with a hot potato in his or her mouth. There have been very few studies analysing the profile and characteristics of the voice changes in tonsillitis or peritonsillitis and none that have compared these changes with those that occur with a hot potato in the oral cavity.'

To remedy this lack of knowledge, the three doctors recruited two sets of volunteers. The first group comprised ten hospital patients whose suffering related to their tonsils. Each volunteer pronounced three particular vowel sounds, which the doctors recorded and subsequently analysed using special software. The second group was ten healthy hospital staffers, 'with each of these participants placing a British new potato of approximately 50 grams in their oral cavity, warmed by microwave to a "hot" but not uncomfortable temperature'.

The doctors detected unmistakable differences. The unique sound of someone burdened with an actual potato, they explain, 'is related to interference with the anterior tongue function from the physical presence of the potato'.

Bhutta, Mahmood F., George A. Worley, and Meredydd L. Harries (2006). '"Hot Potato Voice" in Peritonsillitis: A Misnomer.' *Journal of Voice* 20 (4). 616–22.
Bhutta, Mahmood F., and Harold Maxwell (2008). 'Sneezing Induced by Sexual Ideation or Orgasm: An Under-Reported Phenomenon.' *Journal of the Royal Society of Medicine* 101: 587–91.

DEEP, DARK ROMANCE

Of all the romance books ever written, which has the most surprising depths?

The Romance of Tristan and Iseult? No. *The Romance of Isabel, Lady Burton*? No.

The Romance of Pepperell, Being a Brief Account of the Career of Sir William Pepperell, Soldier, Pioneer, American Merchant and Developer of New England Industry, for Whom the Pepperell Manufacturing Company was Named, and the Towns of Saco and Biddeford in the State of Maine, wherein the First Manufacturing Unit of the Pepperell Company was Established? No.

No, none of those books approach the depths of Charles Elton Blanchard's 1938 thriller, *The Romance of Proctology*.

Blanchard was a proctologist by trade and by temperament. He wrote some twenty books on the subject. *The Romance of Proctology* is his masterpiece.

Later authors were inspired by Blanchard's élan. Emilio de los Ríos Magriñá, for one, is notable for his *Color Atlas of Anorectal Diseases*, published in 1980. But as its title implies, the book lacks romance.

Blanchard pours on the romance. His opening sentence is an irresistible come-on: 'No one knows who was the first doctor to examine the rectal orifice of the human frame.'

The reader grows all quivery as Blanchard shows us history's parade of charismatic proctologists, heroic actions, and frightening tools of the trade.

'These pioneers were earnest seekers after proctological truth', he writes in introducing Dr William Allingham of London. 'Allingham believed in the value of linear cauterization using the Paquelin cautery for proctidentia recti. He claims he was the first (and possibly the last) to insert the whole hand into the rectum.'

The seventeenth-century physician Morgani receives special praise. Blanchard speaks of him on our behalf: 'We are thankful to Morgani that in the midst of all his many researches he, of all the great names at Padua, looked into the human rectum, discovered and named its crypts and pillars.'

'It is strange', Blanchard reminds us, 'how immortality in medicine is often gained by some very minor contribution. Morgani is remembered by the crypts and columns of the rectal outlet. Hilton by his "white line" which is seldom white in the living subject.' He is writing about John Hilton of Guy's Hospital, London – the John Hilton who was known as 'anatomical John' and who was made surgeon to Queen Victoria. Blanchard's reverence for him is nearly boundless: 'I would rather drop one tear on the grave of John Hilton than to place a costly wreath on the tomb of Napoleon.'

Blanchard tips his cap, too, to Dr Joseph M. Mathews of Louisville, Kentucky, of whom he writes: 'Dr Mathews was much like Dr Allingham, jovial, talkative and yet rather sure of his opinions being right. He much preferred to be called "Rectal Specialist" than by any other high-sounding name. To him should go much credit for making proctology a specialty.'

There are, of course, many biological romance books. Anyone who enjoys Blanchard's *The Romance of Proctology* can seek delight also in A. Radclyffe Dugmore's *The Romance of the Beaver*, published in 1914.

Blanchard, Charles Elton (1938). *The Romance of Proctology*. Youngstown, Ohio: Medical Success Press.
Neuhauser, D. (2006). 'Advertising, Ethics and the Competitive Practice of Medicine: Charles Elton Blanchard MD.' *Quality and Safety in Health Care* 15: 74–75.

CATCHING DISCO FEVER

However serious researchers were about discotheques, most of them kept quiet about it for a long time. Then a glorious decade gave birth to two pools of disco studies. One describes injuries, illnesses, and other ills that should or could be blamed on discos and disco music. The other tells about a world of exciting disco-inspired and disco-enabled – in short, disco-fuelled – investigations.

A lone, curious voice, that of M. S. Swani of Birmingham, UK, sounded perhaps the first interested cry. In a letter dated 30 November 1974 published in the *British Medical Journal*, Dr Swani wrote: 'Early deafness in young people as a result of exposure to excessive noise in "discos" must now be assuming epidemic proportions. The importance of this problem has been brought especially to my mind because an 18-year-old medical secretary, who has worked for me, has now been found to be suffering from this condition. If every general practitioner in the country had one such new case a year there would be 20,000 new cases in the country annually.'

Discos became popular in the 1960s and wildly so in the 1970s, but almost no formal disco-themed studies appeared until 1980. Thereafter, disco scholarship flourished.

One stream of reports, perhaps an indirect result of Swani's secretary's what-what-what-ing to her frustrated boss, explained that people who spend too much time listening to much-too-loud music become hard of hearing.

Around the world, doctors published reports raising other medical questions. Among the titles: 'Effect of Discotheque Environment on Epileptic Children' (UK, 1981); 'Acute Central Cervical Cord Injury Due to Disco Dancing' (Ireland, 1983); 'The Dyspeptic Disco Dancer', (Hong Kong, 1988); 'Disco Fever: Epidemic Meningococcal Disease in Northeastern Argentina Associated with Disco Patronage' (Argentina, 1988); and 'Valsalva Retinopathy Associated with Vigorous Dancing in a Discotheque' (Israel, 2007). Roller disco inspired its own sub-genre, with such titles as: 'Roller Disco Neuropathy' (USA, 1981); and 'The Roller Discotheque – A Quickstep to the Hospital? An Analysis of 196 Accidents' (Germany, 1985).

But it wasn't just doctors. Disco opened exciting new worlds for everybody. I will mention just two of the studies that appeared in that breakthrough year, 1980. Margaret Doyle Pappalardo wrote her doctoral thesis, at Boston University in Massachusetts, on 'The Effects of Discotheque Dancing on Selected Physiological and Psychological Parameters of College Students', while graduate student, Bruce Taylor, at the University of Bergen, sought not the side-effects of disco, but its heart.

Taylor's thesis, called 'Shake, Slow, and Selection: An Aspect of the Tradition Process Reflected by Discotheque Dances in Bergen, Norway', appeared in the journal *Ethnomusicology*. He interviewed patrons near the dance floor. 'According to them', Taylor wrote, 'the most important principle is to follow the rhythm and the beat, but variation is also necessary, and a good dancer is interested in the dance as well as in his partner ... Conversations between strangers are begun, personal contact is achieved, and many of the guests who arrived alone are actively interested in leaving for home with a new acquaintance of the opposite sex.'

Even doctors manage, sometimes, to find some delight in disease, especially in describing its effects and after-effects. That seems evident in the phrasing of a case report called 'The University Rollerdisco: An Unusual Cause of a Major Incident', which appeared in the journal *Injury Extra*.

The co-authors play the role of raconteurs rather than traditional, stuffy medicos: 'Rollerdiscos are associated with a high incidence of injury, as is binge drinking. On Valentine's evening 2008, Liverpool University combined these two venerable pastimes at a student event, without informing local health services. Subsequently, emergency services were overwhelmed with Rollerdisco casualties and a "major incident" ... The event itself consisted of a newly laminated hall for roller skating, alcoholic drink promotions and a 1980s themed dress code. Certainly the Accident and Emergency Department became a colourful place with variously injured patients in flamboyant dress and in a generally "exuberant" mood. In all, eight patients were admitted (one patient for every 17 min of the disco).'

CASE REPORT

The University Rollerdisco: An unusual cause of a major incident

A.J. Highcock[a,*], K. Rourke[b], D. Brown[b]

Swani, M. S. (1974). 'Disco Deafness.' *British Medical Journal* 4 (5943): 532.

Peck, R. J., Karen Ng, and Arthur Li (1988). 'The Dyspeptic Disco Dancer.' *British Journal of Radiology* 61 (725): 417–18.

Dewitt, L. D., and H. S. Greenberg (1981). 'Roller Disco Neuropathy.' *Journal of the American Medical Association* 246 (8): 836.

Redmond, J., A. Thompson, and M. Hutchinson (1983). 'Acute Central Cervical Cord Injury Due to Disco Dancing.' *British Medical Journal* 286 (6379): 1704.

Dörner, A., H. J. Kahl, and K. H. Jungbluth (1985). 'The Roller Discotheque: A Quickstep to the Hospital? An Analysis of 196 Accidents.' *Unfallchirurgie* 11 (4): 181–86.

Bar-Sela, S. M., and J. Moisseiev (2007). 'Valsalva Retinopathy Associated with Vigorous Dancing in a Discotheque.' *Ophthalmic Surgery, Lasers & Imaging* 38 (1): 69–71.

Cookson, Susan Temporado, José L. Corrales, José O. Lotero, Mabel Regueira, Norma Binsztein, Michael W. Reeves, Gloria Ajello, and William R. Jarvis (1998). 'Disco Fever: Epidemic Meningococcal Disease in Northeastern Argentina Associated With Disco Patronage.' *Journal of Infectious Diseases* 178 (1): 266–69.

Pappalardo, Margaret Doyle (1980). 'The Effects of Discotheque Dancing on Selected Physiological and Psychological Parameters of College Students.' PhD thesis, Boston University School of Education.

Taylor, Bruce H. (1980). 'Shake, Slow, and Selection: An Aspect of the Tradition Process Reflected by Discotheque Dances in Bergen, Norway.' *Ethnomusicology* 24 (1): 75–84.

Highcock, A. J., K. Rourke, and D. Brown (2008). 'The University Rollerdisco: An Unusual Cause of a Major Incident.' *Injury Extra* 39 (12): 386–88.

IN BRIEF

'PUNK ROCKER'S LUNG: PULMONARY FIBROSIS IN A DRUG SNORTING FIRE-EATER'

by D. R. Buchanan, D. Lamb, and A. Seaton (published in the *British Medical Journal*, 1981)

THE TOOTHLESS RULE OF LOUIS XIV

French teeth are something of a specialty for the president of Great Britain's Royal Historical Society. Colin Jones, who is also a history professor at Queen Mary, University of London, has written two memorable monographs on the subject.

Evidence of a huge Parisian tooth-yanker. 'A French Dentist Shewing a Specimen of His Artificial Teeth and False Palates' by Thomas Rowlandson (1811). Wellcome Library, London

Jones's study, called 'Pulling Teeth in Eighteenth-Century Paris', centres on a literally huge Parisian tooth-yanker called *Le Grand Thomas*. Jones explains: 'For nearly half a century, from the 1710s to the 1750s, Thomas was a standard fixture, a living legend, plying his dental

wares on the Pont-Neuf in Paris ... If the tooth he was attacking re-
pulsed his assaults he would, it was said, make the individual kneel
down, then, with the strength of a bull, lift him three times in the air
with his hand clenched on the recalcitrant tooth.' Jones suggests that
a well-informed toothache sufferer, surveying the major healthcare
options, might reasonably opt for *Le Grand Thomas* or one of his many
self-taught peers.

Surgeons, the people most likely to do a good job, were enjoying a
rise in prestige and fees. They would commonly decline the pedestrian,
relatively low-paying task of tooth-pulling. Doctors and apothecaries
'were both still primarily hands-off practitioners' whose services might
be expensive and whose array of remedies still included things like
'the ingestion of flayed, crushed and cooked mouse'.

Given these alternatives, Jones writes, 'it is not difficult to imagine
that the limited dental skills of the smithy or the therapeutic value of
casseroled mouse must have opened up a niche for a more helpful and
more imaginative approach. This niche appears to have been filled by
men of the stripe of *Le Grand Thomas*'.

Jones also wrote a study called 'The King's Two Teeth'. The title
refers to the two choppers present at birth, in 1638, in the mouth of
Louis XIV, the man who would later be called Louis the Great and
Louis the Sun King. 'To contemporaries', writes Jones, 'this prodigious,
gluttonous, voracious pair of teeth seemed to presage the wonders
which the hungrily devouring prince would in the fullness of time
effect on the map of Europe.'

Jones mentions a tradition of French royal portrait painting: kingly
teeth, even when existing and beautiful, were always hidden behind
closed lips.

But traditions would change.

A much-celebrated portrait done in 1701, of a sixty-three-year-old
Louis 'at the height of his powers' shows the king with impressively
youthful legs and posture. Even with this blatant inaccuracy, Jones says,
'one feature stands out – and shocks – for its stark naturalism: hollow
cheeks and wrinkled mouth reveal a ruler with not a tooth in his head'.

Together with the development of better dental medicine, he concludes, 'the replacement of the tooth puller by the dentist and the emergence on the marketplace of a powerful demand for a different kind of

mouth all in their different ways highlighted a silent revolution of the teeth and the smile which bade to put paid to the Old Régime of Teeth'.

Jones, Colin (2000). 'Pulling Teeth in Eighteenth-Century Paris.' *Past and Present* 166 (1): 100–45.
— (2008). 'The King's Two Teeth.' *History Workshop Journal* 65: 79–95.

MUMMIES' RECIPE FOR GOOD HEALTH

Nowadays, powdered mummy may not be everyone's cup of tea, but for many years it was just what the doctor ordered. That's one of the takeaway messages of Richard Sugg's study '"Good Physic but Bad Food": Early Modern Attitudes to Medicinal Cannibalism and its Suppliers'. Sugg, a research fellow in literature and medicine at Durham University, in the UK, begins with an observation: 'The subject of medicinal cannibalism in mainstream western medicine has received surprisingly little historical attention.'

Sugg tells us that mummy, generally in powdered form, 'having originally been a natural mixture of pitch and asphalt, came in the twelfth century to be associated with preserved Egyptian corpses'. It then 'emerged as a mainstream western medicine' and remained a standard-issue drug until 'opinion began to turn against it in the eighteenth century'.

Physicians prescribed powdered mummy for diverse ailments. An English pharmacopeia published in 1721 specifies two ounces of mummy as the proper amount to make a 'plaster against ruptures'. Ambroise Paré, royal surgeon to sixteenth-century French kings, proclaimed mummy to be 'the very first and last medicine of almost all our practitioners' against bruising.

Dr Paré harboured doubts about the drug's efficacy, lamenting that 'wee are ... compelled both foolishly and cruelly to devoure the mangled and putride particles of the carcasses of the basest people of Egypt, or such as are hanged'. But Paré was an unusually driven doubting Thomas – he lamented having 'tried mummy "an hundred times" without success'.

Sugg's study explains that 'mummy was an important commodity. It is often seen in long lists of merchants' wares and prices.' The marketplace attracted counterfeiters. Sugg supplies an anecdote: 'Tellingly, when Samuel Pepys saw a mummy it was in a merchant's warehouse;

while "the abuses of mummy dealers in selling inferior wares" were especially widespread and notorious by the end of the seventeenth century.'

The best suppliers maintained high standards. The presumably admirable recipe used by seventeenth-century German pharmacologist Johann Schroeder included 'the cadaver of a reddish man (because in such a man the blood is believed lighter and so the flesh is better), whole, fresh without blemish, of around twenty-four years of age, dead of a violent death (not of illness), exposed to the moon's rays for one day and night, but with a clear sky. Cut the muscular flesh of this man and sprinkle it with powder of myrrh and at least a little bit of aloe, then soak it.'

Sugg, Richard (2006). "'Good Physic but Bad Food": Early Modern Attitudes to Medicinal Cannibalism and its Suppliers.' *Social History of Medicine* 19 (2): 225–40.

TO BE MICHAEL JACKSON, AFTER A REFASHIONING

Mandibular angle augmentation with the use of distraction and homologous lyophilized cartilage in a case of morphing to Michael Jackson surgery

M.Y. Mommaerts [1*], J.S.V. Abeloos [1], H. Gropp [2]

Summary

Correction of an ill-defined mandibular angle is not an easy task, whether it is requested by the "congenital, orthognathic or cosmetic" patient. Deliberate over-correction has not been reported to our knowledge. This

In 1997, a twenty-four-year-old Belgian man requested that his head be reconstructed to make him resemble the singer Michael Jackson. Three plastic surgeons granted his wish. Their report about it, published in the journal *Annales de Chirurgie Plastique et Esthetique*, is a lovely sight to behold. The loveliness is partly in the detailed technical description, monochromatically in the set of before-and-after X-rays of the facial bones, and memorably in the medically stylish photographs that show the young man before and after his course of treatment.

The doctors, Maurice Mommaerts and Johan Abeloos, of the Hôpital Général Saint-Jean in Bruges, Belgium, and H. Gropp, of the Diakoniehospital in Bremen, Germany, described the patient's challenge to them: 'His quest was to obtain the facial features of Michael Jackson, his idol that he imitated professionally.' This was an unusual demand. The doctors explain that 'normally, patients strive for an ideal, beautiful, normal contour [of the facial bones]. We were confronted with a patient who requested a three-dimensional overcorrection.'

Their patient was no ordinary young man. He impressed the doctors with the firmness of his desire, but also with his detailed knowledge of his own craniofacial anatomy (especially his gonial angles and malar prominence).

This task, the doctors decided, after only minimal hesitation, was something they could do. 'After thorough discussion and psychiatric analysis, we agreed to morph him in a way that all changes could be undone and that the tissues were not at risk for considerable permanent damage.'

The case was both easy and hard. The surgeons immediately saw simple ways to rearrange the young man's chin and also his cheekbone arches. But how to achieve the desired posterior-mandibular augmentation? That was the puzzler; solving it would be a medical first.

The doctors rose to the posterior-mandibular augmentation challenge. They conquered it and, in so doing, made history. Two rounds of surgery did the trick. Full details are in their report. For non-specialists, the important feature may be the simple and comforting piece of knowledge: Yes, we now know, it *is* possible to surgically morph a long-jawed Belgian white youth so that he looks just like Michael Jackson.

Yet, a prominent institution that houses that particular type of individual suddenly has, at least potentially, a big problem. Hordes of people want to see him, touch him, admire him, maybe even serve legal papers on him. I found no reports of that happening with this Belgian doppelganger. I suspect that is because the surgeons kept up with the medical literature, and had learned from a 1996 report in the journal *Hospital Security and Safety Management*. That instructional article, written in the wake of Mr Jackson's unfortunate and dramatic collapse on stage in New York City, is called: 'Michael Jackson at Beth Israel: Handling Press, Fans, Gawking Employees'.

Mommaerts, M. Y., J. S. Abeloos, and H. Gropp (2001). 'Mandibular Angle Augmentation with the Use of Distraction and Homologous Lyophilized Cartilage in a Case of Morphing to Michael Jackson Surgery.' *Annales de Chirurgie Plastique et Esthetique* 46 (4): 336–40.

N. A. (1996). 'Michael Jackson at Beth Israel: Handling Press, Fans, Gawking Employees.' *Hospital Security and Safety Management* 16 (12): 10–11.

PURSUING A WRETCHED ITCH

Can capsaicin – the chemical that causes most of the burning sensation when you chomp on a chilli pepper – relieve itching at the nether
end of the digestive tract? A team of Israeli scientists tried to find out.

They tackled a maddening medical condition called 'idiopathic
intractable pruritus ani'. Most people, including most doctors when
they are talking informally to each other, use the less-formal name:
'persistent butt itch'. It is one of a wide class of medical conditions that
sound humorous until you experience them yourself. And then they
still sound funny, which perhaps adds to the discomfort.

Dr Eran Goldin and a large team of colleagues at Hadassah University Hospital, in Jerusalem, collected forty-four patients who suffered from chronic butt itch. Each had endured at least three months
of suffering. None had responded to the traditional treatments – gentle
washing and drying of the affected area, and avoidance of certain
foods that are famous for causing chronic butt itch.

Coffee, tea, cola, beer, chocolate, and tomatoes are thought to be
the six biggest causes of the problem, identified as such in a 1997 report
by William G. Friend of the University of Washington. Friend believed
that coffee is the main culprit, responsible for about eighty percent of
all cases of intractable butt itch. Drink less coffee and you'll be able to
sit still, if you are one of the luckier butt itch sufferers. The forty-four
Israeli itch victims, though, did not have that sort of luck. Theirs was
an itch of unknown origin, a head-scratching puzzle for any doctor
who tried to treat them.

Goldin and his team solved this puzzle for thirty-one of their
forty-four patients by applying the capsaicin topically. Four patients
did feel what Goldin called 'a very mild perianal burning lasting
10–15 minutes' after the treatment, but apparently this was for them
a small price to pay.

Some months later, the doctors checked up on eighteen of the
patients. All said they were still feeling pretty good, so long as they
gave themselves an anal dose of capsaicin every day or two. The Goldin
report concluded that 'capsaicin is a new, safe, and highly effective
treatment for severe intractable idiopathic pruritus ani'.

While new for treating this very specific ailment, capsaicin was already, as the doctors themselves point out, generally 'known to be effective and safe in the treatment of pain and itching'. Capsaicin was also, of course, known to have rather ferocious effects when placed into the front end of a person's digestive system.

A 2002 experiment by doctors at L. Nair Hospital in Mumbai, India, explored both sides of the action. The research team fed ten grams of red chilli powder (in other words, a heaping dose of capsaicin) to twenty-one men who have well-tempered bowels. The doctors report that this 'increases the rectal threshold for pain'. You will forgive me, I hope, for not describing how they performed that measurement.

Lysy, J., M. Sistiery-Ittah, Y. Israelit, A. Shmueli, N. Strauss-Liviatan, V. Mindrul, D. Keret, and E. Goldin (2003). 'Topical Capsaicin – A Novel and Effective Treatment for Idiopathic Intractable Pruritus Ani: A Randomised, Placebo Controlled, Crossover Study.' *Gut* 52: 1323–26.

Friend, William G. (1977). 'The Cause and Treatment of Idiopathic Pruritus Ani.' *Diseases of the Colon and Rectum* 20 (1): 40–42.

Agarwal, M. K., S. J. Bhatia, S. A. Desai, U. Bhure, and S. Melgiri (2002). 'Effect of Red Chillies on Small Bowel and Colonic Transit and Rectal Sensitivity in Men with Irritable Bowel Syndrome.' *Indian Journal of Gastroenterology* 21 (5): 179–82.

THE FINGERNAILS OF DR BEAN

Many people, especially academics and taxi drivers, take pride in having arcane knowledge at their fingertips. Dr William B. Bean bested them all. Bean's arcane knowledge was not only at his fingertips; it was about them. Bean spent much of his adult life monitoring the growth of his fingernails. He trimmed his nails neither to be fashionable nor to add to his press clippings. He did it for science.

William B. Bean (born 1909, died 1989) conducted what is known as a longitudinal self-study of fingernail growth. It is one of the few such studies known, and perhaps the lengthiest. Bean taught for many years at the University of Iowa College of Medicine and later at the University of Texas medical branch at Galveston, Texas. The cuticle research was published, in pieces and at intervals, in the *Archives of Internal Medicine*, of which Bean happened to be editor.

In 1968, the first of the Bean nail papers arrived in print. Called 'Nail Growth: Twenty-Five Years' Observation', its timing was unfortunate for Bean, in that the world was distracted by riots, assassinations, the Vietnam war, and the nail-biting American presidential election in which Richard Nixon rose to power. The year 1974 saw the publication of Bean's extended observations. His paper 'Nail Growth: 30 Years of Observation' was published just a few weeks after President Nixon's attention-grabbing resignation from the American presidency. Again, Bean received scant acclaim.

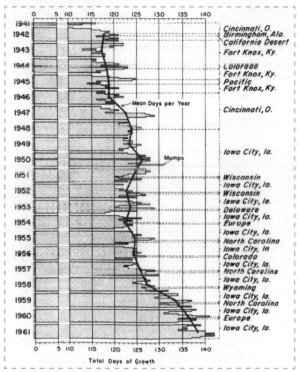

Dr Bean's twenty-year nail chart from 'Nail Growth and Unusual Fingernails'

Two years later, perhaps growing a little impatient, Bean drummed his metaphorical fingertips on a different tabletop, publishing a cuticle-centric essay not in his own journal, but in the *International Journal of Dermatology*. Under the headline 'Some Notes of an Aging Nail Watcher', he explained: 'Growth of deciduous tissues gives us a natural kymograph to record secular trends and in some instances makes the mark on the moving record. For the observant clinician, knowledge of the rate of nail growth may permit an occasional spectacular diagnosis, although much more often it merely adds a small bit to our understanding of simple but basic biological principles in health and disease.' This seems to have produced a gratifying response.

Thereafter, Bean returned to his original, deliberate publication schedule. In 1980 he produced 'Nail Growth: Thirty-Five Years of Observation'. It is as complete a story as the world has ever seen about the growth of one physician's fingernails. Here is his summary: 'A 35-year observation of the growth of my nails indicates the slowing of growth with increasing age. The average daily growth of the left thumbnail, for instance, has varied from 0.123 millimeters a day during the first part of the study when I was 32 years of age to 0.095 millimeters a day at the age of 67.'

Bean, William B. (1962). 'A Discourse on Nail Growth and Unusual Fingernails.' *Transactions of the American Clinical and Climatological Association* 74: 152–67.

— (1968). 'Nail Growth: Twenty-Five Years' Observation.' *Archives of Internal Medicine* 122 (4): 359–61.

— (1974). 'Nail Growth: 30 Years of Observation.' *Archives of Internal Medicine* 134 (3): 497–502.

— (1976). 'Some Notes of an Aging Nail Watcher.' *International Journal of Dermatology* 15 (3): 225–30.

— (1980). 'Nail Growth. Thirty-Five Years of Observation.' *Archives of Internal Medicine* 140 (1): 73–76.

NEW PET THEORY

Some Australian researchers have put forward a new pet theory about older people and their beloved pets, which many have claimed, including the *Medical Journal of Australia*, are 'good for health' – the health of humans. This new theory is blunt in overturning this assumption.

Susan Kurrle and Robert Day, of the Hornsby Ku-ring-gai Health Service in Sydney, Australia, and Ian Cameron, of the University of Sydney, looked at cases of pet-related falls that brought patients seventy-five years and older to one particular hospital during a six-month period. They defined pets as 'an animal which is kept as a companion and is treated with affection'. This included animals such as goats and donkeys, as well as dogs, cats, and birds. They narrowed their definition of the injured to fall victims who sustained a traumatic bone fracture. Their analysis excluded injuries that 'occurred as a result of older people being startled by mice, cockroaches or spiders, as these animals were not considered pets for the purpose of this study.'

The circumstances of each case, as presented in the report, are plaintively stark. Here are a few, each quite typical:

1) Taking Jack Russell terrier for walk using retractable leash. Dog ran round and round patient's legs and pulled him over.

2) Climbing stile over fence to feed mohair goats, slipped and fell to ground.

3) Feeding donkey from bucket. Donkey nudged patient, pushing her over backwards.

4) Slipped on puddle of urine from new Labrador pup. Fell against wooden arm of armchair.

5) Fell forwards while trying to prevent young puppy from diving into fish tank.

6) Fell sideways in garden while trying to stop cat catching a blue tongue lizard.

7) Tripped over black cat in darkened hallway.

8) Fall while attempting to move quickly out back door as cat carried live snake in through side door.

'There were no deaths recorded as a result of the fall-related fractures', Kurrle et al. tell us, 'but one of the animals involved (a cat) died when its owner fell and landed on it.'

Kurrle, Susan E., Robert Day, and Ian D. Cameron (2004). 'The Perils of Pet Ownership: A New Fall-Injury Risk Factor.' *Medical Journal of Australia* 181: 682–83.

ADDRESSING THE KARAOKE PANDEMIC, VOCALLY

A scientific experiment may look like torture, and sound like torture, yet still be free of legal ramifications. At the University of Hong Kong, Edwin M.-L. Yiu and Rainy M. M. Chan did an experiment that smacks of torture for the participants, the experimenters, and anyone within earshot. Their published report has a title that evokes wretchedness: 'Effect of Hydration and Vocal Rest on the Vocal Fatigue in Amateur Karaoke Singers'.

The experiment brought several hours of continuous, mounting, painful discomfort to a group of human volunteers. Yet the scientists' aim was noble. They write that 'karaoke singing is a very popular entertainment among young people in Asia ... It is not uncommon to find participants singing continuously for four to five hours each time. As most of the karaoke singers have no formal training in singing, these amateur singers are more vulnerable to developing voice problems under these intensive singing activities.'

This modestly understates the problem. Many thousands of young persons sing karaoke. Multiply that by the duration of singing – four or five hours. Now multiply that by the average number of times per week each person sings karaoke. Then multiply by fifty-two weeks. The resultant sum represents a groaning annual burden of painful singing, on a continental scale. And that's just Asia. Karaoke is pandemic on at least six continents.

The experimental subjects were a carefully chosen bunch, all in their early twenties, in good health, and in the habit of singing karaoke at least twice a week. They had no formal voice or singing training, no history of voice problems, and no chronic psychiatric problems worth mentioning.

Yiu and Chan performed this experiment at the university's voice research laboratory. Each person 'was asked to sing in a quiet room with karaoke facility, which provided music video on a television and background music with echo effects ... The participants were required to sing continuously until they reported feeling fatigue with their voices and could not sing anymore.'

Effect of Hydration and Vocal Rest on the Vocal Fatigue in Amateur Karaoke Singers

Edwin M-L Yiu and Rainy MM Chan

Hong Kong, China

Ten of them got to rest for a minute after each song, and drink some water. The other ten received neither hydration nor rest; they bopped till they dropped, so to speak.

The hydrated singers sang longer, if not better, than those denied liquid. The former averaged more than one hundred minutes of warbling, the latter about eighty-five. (The four to five hours they claim to sing while at karaoke clubs presumably includes lots of down time.)

Yiu and Chan did find a surprise. They had expected the wet singers to sing better than the dry. This expectation was largely unmet. Assessments by trained ears and eyes (the latter involving phonetograms – electroacoustically produced graphs of pitch and loudness) showed that, warble for warble, vocal quality levels were roughly the same for both groups.

Unskilled singers, one might infer from this, seldom exceed mediocrity yet seldom fail to achieve it. Occasional water and rest can help them prolong their remarkable record of achievement.

Yiu, Edwin M.-L., and Rainy M. M. Chan (2003). 'Effect of Hydration and Vocal Rest on the Vocal Fatigue in Amateur Karaoke Singers.' *Journal of Voice* 17 (2): 216–27.

HIVES ON THE PITCH

'This is the first reported case of an urticarial rash apparently caused by the frustration of watching England play football.'

With these words, written in 1987, a London general practitioner trainee named P. Merry alerted readers of the *Journal of the Royal Society of Medicine* to a little-suspected risk of rooting for a World Cup football team. Rooting can cause emotional upset, which can cause urticaria. Urticaria is also known as 'hives'.

Here's what happened to the patient, who followed the game on TV: 'When Portugal scored the only goal of the match to win 1-0, he became extremely upset, and developed the rash of urticaria affecting his trunk and limbs. This persisted for 36 hours and then settled.' Four days later, the man watched England vs. Morocco. 'When a member of the English team was sent off, he became agitated and subsequently developed the same rash of urticaria on his trunk and limbs.'

Then, in 2006, thirty-four-year-old Paul Hucker, of Ipswich, Suffolk, UK, made headlines because he bought an insurance policy against possible trauma brought on by an England defeat in the World Cup. Many chuckled at the news. A wander through the medical literature suggests the chucklers should temper their amusement.

Five researchers at the University of Bristol published a warning in 2002, in the *BMJ*, that 'myocardial infarction can be triggered by emotional upset, such as watching your football team lose an important match'. Their main evidence: British hospital statistics accumulated at the time of the 1998 World Cup. 'Risk of admission for acute myocardial infarction', the doctors point out, 'increased by 25% on 30 June 1998 [the day England lost to Argentina in a penalty shoot-out] and the following two days.'

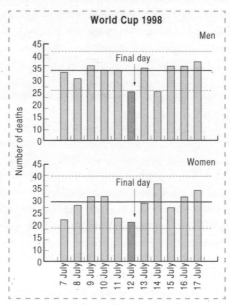

Number of deaths from myocardial infarction, French men vs. French women. France played Brazil in the final of the World Cup on 12 July 1998.

Four researchers in Lausanne, Switzerland, say a similar thing happened during the 2002 World Cup. They lay out their stats in a 2006 issue of the *International Journal of Cardiology*. The number of sudden cardiac deaths was sixty-three percent higher during the World Cup than during the equivalent period a year earlier, when there was no World Cup competition. The doctors try to analyse it: 'We explain this by an increase in mental stress and anger and possible unhealthy behaviour (increased alcohol and tobacco consumption, decreased medical compliance) of football supporters. The lethal effect of mental stress and anger has been attributed to its activation of the sympathetic nervous system leading to hypertension, impaired myocardial perfusion in the setting of atherosclerotic disease and a high degree of cardiac electrical instability precipitating malignant arrhythmias.'

Fandom carries danger, yes, but there's a special payoff for those whose side does capture the ultimate glory. Or so implies a study that appeared in 2003 in the journal *Heart*. Written by two French doctors, the title proclaims: 'Lower Myocardial Infarction Mortality in French Men the Day France Won the 1998 World Cup of Football'.

Merry, P. (1987), 'World Cup Urticaria.' *Journal of the Royal Society of Medicine* 80 (12). 779.

Carroll, D., S. Ebrahim, K. Tilling, J. Macleod, and G. D. Smith (2002). 'Admissions for Myocardial Infarction and World Cup Football: Database Survey.' *BMJ* 325: 1439–42.

Katz, Eugène, Jacques-Thierry Metzger, Alfio Marazzi, and Lukas Kappenberger (2006). 'Increase of Sudden Cardiac Deaths in Switzerland during the 2002 FIFA World Cup.' *International Journal of Cardiology* 107 (1): 132–33.

Berthier, F., and F. Boulay (2003). 'Lower Myocardial Infarction Mortality in French Men the Day France Won the 1998 World Cup of Football.' *Heart* 89 (3): pp. 555–56.

OBJECT RCSHC/P 192

The rectum of the Bishop of Durham sits on display in London, awaiting your examination. No longer attached to the bishop, it rests alone inside a glass jar in the Hunterian Museum at the Royal College of Surgeons of England. The museum calls it by the formal name: Object RCSHC/P 192.

Visitors can casually admire the object's beauty. Scholars and poets can find unexpected delights in studying and writing up the bishop's rectum. This apparently humble body part can boast a historic connection to John Hunter, the surgeon whose collection of medical memorabilia eventually grew to become the Hunterian Museum.

The museum officially gives a simple description of Object RCSHC/P 192: 'A rectum showing the effects of both haemorrhoids and bowel cancer. The patient in this case was Thomas Thurlow (1737–91), the Bishop of Durham. Thurlow had suffered for some time from a bowel complaint, which he initially thought was the result of piles. He consulted John Hunter after a number of other physicians and surgeons had failed to provide him with a satisfactory diagnosis. Hunter successfully identified the tumour through rectal examination but recognised that it was incurable. Thurlow died ten months later.'

Hunter wrote extensive notes about how he entered the case, examined the rectum (which at the time was, of course, still an integral part of the bishop), and immediately recognized, by feel, that it had an incurable tumour.

The notes also tell how events played out. The bishop, disbelieving Dr Hunter's diagnosis, then tried to cure himself with a nostrum called Ward's White Drops. He was choosing to rely on past experience with a lesser ailment, rather than accept Hunter's professional assessment. Hunter notes that 'his Lordship had, about ten years ago, the piles, for which he took Ward's Paste, and was cured'.

The White Drops did not cure the bishop's cancer. Instead, his discomfort increased. Hunter writes that the family then called in 'Taylor the cattle-doctor to attend him, and I was asked to examine this doctor, to see whether it was likely he should do mischief or not'. Hunter concluded that Taylor would do no mischief. Taylor deferred happily to the renowned physician's opinions and, with his approval, gave the bishop opium and ointments, to ease the distress.

Ten months later, the bishop breathed his last. John Hunter performed an autopsy, savouring the opportunity to write a detailed technical assessment of the tumour and of its role in killing a patient who doubted the doctor's diagnosis.

The copious details are a bit grisly for a general audience. Hunter's notes were intended for himself or for others of his profession, should he or they encounter a similar rectum or a similar patient. Now, more than two hundred years later, the story, and a good view of the rectum, are available to anyone who seeks enlightenment.

Steve Farrar noticed the bishop's rectum and brought me to visit it. That resulted in tea with Simon Chaplin, the museum's director, who has a special fondness for and knowledge of historic body parts. I am and will eternally be grateful to both men for their insights into the remaining bit of bishop.

MAY WE RECOMMEND

'NO-SCALPEL VASECTOMY AT THE KING'S BIRTHDAY VASECTOMY FESTIVAL'

by Apichart Nirapathpongporn, Douglas H. Huber, and John N. Krieger (published in the *Lancet*, 1990)

IOI USES FOR THE SACRED FORESKIN

A study called 'The Circumcision of Jesus Christ' pioneers a new flavour of interdisciplinary research: urology at last joins forces with theology. Published in the *Journal of Urology*, the study focuses on what happened to Jesus's foreskin during and especially after biblical times.

Lead author Johan J. Mattelaer brings a broad perspective to this narrow topic. A past chairman of the History Office of the European Association of Urology in Kortrijk, Belgium, and professor emeritus of psychiatry at the University of British Columbia in Vancouver, Mattelaer earlier wrote a book called *The Phallus in Art and Culture*. And shortly before taking on the sacred foreskin project, he teamed up with Austrian-Canadian neuropsychologist Wolfgang Jilek to write a study called 'Koro: The Psychological Disappearance of the Penis'. For the Jesus circumcision study, Mattelaer and colleagues Robert A. Schipper and Sakti Das delved into two thousand years' worth of religio-phallocentric writings, paintings, sculpture, music, and theological disputes.

There is art aplenty, they explain, but 'it seems paradoxical that uncircumcised Christian artists created so many images relating to the circumcision of Jesus in painting and sculpture. In Belgium alone there are no less than fifty-four listed works in churches, museums, and public buildings relating to Christ's circumcision, including paintings, grisaille, frescos, statues, altarpieces, stained glass windows and keystones.' Greek and Russian Orthodox church icons, they report, commonly contain circumcision images.

Musicians produced only a few works. The most prominent is 'Missa Circumcisionis Domini Nostri Jesu Christi' ('Mass for the Circumcision of Our Lord Jesus Christ'), composed by Jan Dismas Zelenka of Dresden in 1728.

Churches, museums, crusaders, and kings sought to have and hold the actual foreskin. The study notes that 'the Dominican scholar A. V. Müller, writing in 1907, could list no fewer than 13 separate locations, all of which claimed to possess the sacred foreskin as their holiest relic. We have been able to extend this list to 21 churches and abbeys, which at one time or another are reputed to have held Christ's foreskin.'

The study also reports that King Henry V stole the genuine article – the one so identified by Pope Clement VII – from the French in 1422, and that 'the monks of Chartres were only able to recover it with great difficulty'.

Several theologians devoted their lives to the foreskin. Two remain emblematic. St Catherine of Siena (1347–80), to symbolize her marriage with Christ, 'was reputed to wear the foreskin of Jesus as a ring on her finger'. A generation or so earlier, the Austrian nun Agnes Blannbekin 'led a life devoted to the foreskin of Jesus'. The study says: 'She was obsessed by the loss of blood and the pain which the redeemer had suffered during his circumcision. On one occasion when she was moved to tears by the thought of this suffering, she suddenly felt the foreskin on her tongue.'

The study reproduces a 1523 painting of St Catherine and her ring, but, perhaps deferring to current tastes, supplies no visual image of Agnes Blannbekin.

Mattelaer, Johan J. (2003). *The Phallus in Art and Culture*. Arnhem, The Netherlands: European Association of Urology History Office.
—, Robert A. Schipper, and Sakti Das (2007). 'The Circumcision of Jesus Christ.' *Journal of Urology* 178: 31–34.
—, and Wolfgang Jilek (2007). 'Koro: The Psychological Disappearance of the Penis.' *Journal of Sexual Medicine* 4 (5): 1509–15.

DEATH BY ASPIRATION

One's aspirations can kill – if Dr Sakae Inouye, of Otsuma Women's University in Tokyo, is correct – and Chinese aspirations are particularly deadly.

Inouye devised a simple theory about a vexing public health problem. His theory is this: the English language, when spoken by someone who normally speaks the Chinese language, can be lethal.

Inouye drove his train of logic through the pages of the *Lancet*: 'Severe acute respiratory syndrome (SARS) is transmitted via droplets spread by infected individuals. Droplets are generated when patients cough and, to a lesser extent, when they talk during the early stages of disease. I believe that the efficiency of transmission of SARS by talking might be affected by the language spoken.'

Here are the details of Inouye's reasoning. They are subtle. They are breathtaking. They should perhaps be read silently.

- The disease called SARS seems to have originated in China.
- China has had millions of visitors from the US, and even more visitors from Japan.
- Some American visitors (about seventy out of 2.3 million) got the disease – but *no* Japanese visitors did.
- There must be a reason for that.
- The reason must be: language. In both Chinese and English, many sounds have a strong accompanying exhalation of breath – but Japanese has no such sounds.
- The final step in the chain brings these pieces together. It is frightful. Dr Inouye writes that. 'A Chinese attendant in a souvenir shop probably speaks to American tourists in English, and to Japanese tourists in Japanese. If the shop assistant is in the early stages of SARS and has no cough, I believe American tourists would, hence, be exposed to the infectious droplets to a greater extent than would Japanese tourists.'

Inouye does not specify a particular dialect of Chinese, so at the moment all are suspect.

If one's spoken language is dangerous, can it be altered? Nearly a century ago, future Nobel Prize winner George Bernard Shaw raised this very question. In the printed preface to his play *Pygmalion*, about a professor who painstakingly alters the speech patterns of a young woman, Shaw wrote: 'The change wrought by Professor Higgins in the flower girl is neither impossible nor uncommon ... But the thing has to be done scientifically, or the last state of the aspirant may be worse than the first.'

Inouye, Sakae (2003). 'SARS Transmission. Language and Droplet Production.' *Lancet* 362 (9378): 170.

IN BRIEF

'ATTEMPTED SUICIDE OR HITTING THE NAIL ON THE HEAD: CASE REPORT'

by A. S. Spears (published in the *Journal of the Florida Medical Association*, 1994)

The authors at H. Lee Moffitt Cancer Center & Research Institute, Tampa, Florida, report: 'A case is reported of attempted suicide by hammering nails through the skull into the brain. This unique attempt at self-destruction was unsuccessful and the treatment, initially by an untrained first-aider and then by a neurosurgeon, was surprisingly simple.'

NINE

SERIOUSLY DEADLY

IN BRIEF

'A PARTIALLY MUMMIFIED CORPSE WITH PINK TEETH AND PINK NAILS'

by C. Ortmann and A. DuChesne (published in the *International Journal of Legal Medicine*, 1998)

Some of what's in this chapter: The man who studies sick jokes • Christian End and some dead sports fans • Complications involving brown tree snakes, poisoning, and a parachute • A brief history of certain necrophilia laws • TIT, GAS, and variations of the F-word • Valuing of dead artists • Digging around in churches • The screw-in coffin • The knife muncher

THE BIRTHING OF SICK JOKES

Alan Dundes liked to study uncomfortable jokes and the people who tell them. As his 1979 study called 'The Dead Baby Joke Cycle', published in the journal *Western Folklore*, explains: 'Dead baby jokes are not for the squeamish or the faint of heart. They are told mostly by American adolescents of both sexes in joke-telling sessions with the intent to shock or disgust listeners. "Oh how gross!" is a common (and evidently desired) response to a dead baby joke. Teenage informants of the 1960s and 1970s indicate that dead baby jokes were often used in a "gross out" in which each participant tries to outdo previous joketellers in recounting unsavory or crude folkloristic items.'

To Dundes, when a large group of people persistently make uncomfortable jokes about something, it's something they are uncomfortable about. Thus, he writes that dead baby jokes are popular in the US because of 'the traditional failure of Americans to discuss disease and death openly ... many Americans prefer not to say that an individual is dead or has died.'

Dundes, a longtime professor of anthropology at the University of California, Berkeley, is himself dead, having entered that state in 2005.

He appreciatively blamed England for introducing 'sick humour' to the US, arguing that probably the American variety 'was inspired by a minor English poet Harry Graham, who specialized in light verse and amusing doggerel. In 1899 he published *Ruthless Rhymes for Heartless Homes* and one rhyme in this volume ran as follows:

> *Billy, in one of his nice new sashes*
> *Fell in the fire and was burnt to ashes;*
> *Now, although the room grows chilly,*
> *I haven't the heart to poke poor Billy.'*

In another study, 'Polish Pope Jokes', Dundes presents samples representative of many different varieties of Polish pope jokes, and remarks: 'It was probably inevitable that the Polish-Americans' hope that the election of a Polish Pope would curtail or contain the Polish joke cycle would be in vain. Quite the opposite occurred. The election provided a fresh impetus for a new burst of creativity in the cycle.'

A Dundes' monograph called 'Six Inches from the Presidency: The Gary Hart Jokes as Public Opinion' examines the joke cycle touched off by the withered candidacy of Gary Hart, the front-running Democratic party candidate for the 1988 US presidential election. The joke frenzy began when newspapers published photographs of Hart, in the absence of Mrs Hart, installing a young actress on his lap during an overnight trip 'from Miami to Bimini on a boat with the unlikely but apt name of "Monkey Business"'.

Dundes' best-known book is called *Life is Like a Chicken Coop Ladder: A Portrait of German Culture Through Folklore*. It explores the many variants of the German proverb 'life is like a chicken coop ladder – shitty from top to bottom'. In 174 pages, Dundes plumbed the anal/erotic nature of German culture, and presented evidence for his thesis that Teutonic parents' overemphasis on cleanliness gives their children a lifelong love of scatological humour and imagery.

Dundes, Alan (1979). 'The Dead Baby Joke Cycle.' *Western Folklore* 38 (3): 145–57.

— (1979). 'Polish Pope Jokes.' *Journal of American Folklore* 92 (364): 219–22.

— (1989). 'Six Inches from the Presidency: The Gary Hart Jokes as Public Opinion.' *Western Folklore* 48 (1): 43–51.

— (1984). *Life is Like a Chicken Coop Ladder: A Portrait of German Culture Through Folklore*. New York: Columbia University Press.

CHRISTIAN END INVESTIGATIONS

Despite being blessed with a colourful name, Professor Christian End demonstrates that one can make significant discoveries by looking at unglamorous questions.

End, based at Xavier University in Cincinnati, Ohio, specializes in probing the psychology of sports fans. In the year 2009 he blew past his professional competitors, who generally confine their interest to the living, when he published a study in the journal *Perceptual and Motor Skills* called 'Sport Fan Identification in Obituaries'.

End and three colleagues examined 1101 obituaries in nineteen American and Canadian newspapers. For each, they noted whether or not the deceased was identified as being a sports fan. (They give this example of a clear indicator: 'She was a fan of the Red Sox'.) And they noted whether the individual was a man or a woman.

The End team was testing a novel theory. 'It was hypothesized', they write, 'that a greater proportion of men's obituaries than women's would mention the deceased individual's sport fan identification.'

They learned that twenty-four percent of the dead males were celebrated postmortem as being sports fans, but only 7.7% of the women were accorded that distinction. Thus, End and his co-authors, report, their hypothesis was proved correct.

End sub-specializes in a rather different aspect of sports fandom psychology. It is epitomized by his 2003 monograph (done with a different four collaborators) called 'Perceptions of Sport Fans Who BIRG'. BIRG, the study explains for readers not familiar with this branch of psychology, is an acronym for 'basking in reflected glory'. The End oeuvre includes at least two other published studies that delve into the multidimensional puzzle of sports fandom BIRG.

The inquiries of this Christian End extend far beyond sports fandom.

In a 2007 study entitled 'Unrealistic Optimism in Internet Events', published in the journal *Computers in Human Behavior*, he and another three collaborators 'assessed the tendency for individuals to be unrealistically optimistic about internet related activities' such as downloading music, 'using maps', 'finding a bargain', and 'finding a sought item'. The report's major discovery was that 'heavy internet users' are more optimistic than 'light users' about succeeding at those tasks.

In 2010, End broke new ground. Working with yet another combination of three colleagues (one of whom, Shaye Worthman, also laboured on the obituary study), he published 'Costly Cell Phones: The Impact of Cell Phone Rings on Academic Performance'. For the study, university students were asked to watch a video and take notes. Then the researchers tested the students about the video, and evaluated their notes. For some individuals, the video session 'was disrupted by a ringing cell phone'. Those students (1) 'performed significantly worse' on the test than the ones who were not interrupted, and (2) took crappy notes.

Thus came the researchers to their great discovery. In their words: 'The hypothesis that the cell phone rings would impair performance was confirmed.'

End, Christian M., Jeffrey L. Meinert Jr, Shaye S. Worthman, and Gregory J. Mauntel (2009). 'Sport Fan Identification in Obituaries.' *Perceptual and Motor Skills* 109 (2): 551–54.

End, Christian M., Beth Dietz-Uhler, N. Demakakos, M. Grantz, and J. Biaviano (2003). 'Perceptions of sport fans who BIRG', International Sports Journal 7, 139-149.

Davis, M., and Christian M. End (2005). 'The Economic Impact of Basking in the Reflected Glory of a Super Bowl Victory.' International Association of Sports Economists Conference Papers, http://ideas.repec.org/p/spe/cpaper/0524.html.

Dietz-Uhler, Beth, Elizabeth A. Harrick, Christian End, and Lindy Jacquemotte (2000). 'Sex Differences in Sport Fan Behavior and Reasons for Being a Sport Fan.' *Sport Behavior* 23 (3): 219–30.

Campbell, Jamonn, Nathan Greenauer, Kristin Macaluso, and Christian End (2007). 'Unrealistic Optimism in Internet Events.' *Computers in Human Behavior* 23 (3): 1273–84.

End, Christian M., Shaye Worthman, Mary Bridget Mathews, and Katharina Wetterau (2010). 'Costly Cell Phones: The Impact of Cell Phone Rings on Academic Performance.' *Teaching of Psychology* 37 (1): 55–57.

PERFECTING THE DROP-DEAD-MICE SYSTEM

If you're going to lace dead mice with poison, and drop them from helicopters into a rainforest in Guam in such a way that they become entangled high in the trees, where they might murder the brown tree snakes, but you want to avoid (as much as possible) having the toxically tasty mouse corpses fall all the way to the ground where they could instead get gobbled by coconut crabs, perhaps you should graft them onto something like a parachute. Peter Savarie, Tom Mathies, and Kathleen Fagerstone of the National Wildlife Research Center in Fort Collins, Colorado, did just that. At a symposium in 2007, they told all about it in a report called 'Flotation Materials for Aerial Delivery of Acetaminophen Toxic Baits to Brown Tree Snakes'.

Tree snakes have lived in Guam only since the late 1940s, explain Savarie, Mathies, and Fagerstone. Critics maintain that the snakes have: eaten to near extinction some native birds, lizards, and fruit bats; preyed on poultry; bitten small children; and 'cause[d] power outages by climbing on electrical transmission wires'. Thus came a clamour to get rid of the snakes.

The most obvious way to do that, as certain biologists see it, is to get dead mice, 'treat' them with acetaminophen, stuff the tempting acetaminophen/mouse treats in PVC tubes, and put those where the snakes are. 'However', complains the report, 'PVC tubes are not practical for delivery of baits to remote areas of jungle or the forest canopy. Further, it is important that baits entangle in the canopy and not fall to the ground where they can be scavenged by non-target animals such as crabs.'

Shortly past the turn of the century came an innovation: small parachutes 'hand dropped from a helicopter have been used as flotation devices for entangling dead mice in the forest canopy'. Those early tests used parachutes made of either plastic or corn starch – but the one can take years to biodegrade, and the other dissolves too quickly in the wet.

Savarie, Mathies, and Fagerstone tried several alternatives.

In a trial run, thawed frozen dead mice 'attached to biodegradable jute netting by a 30.5 centimeter-long cotton thread to a rear leg were deployed by hand from a US Navy Knighthawk MH-60S helicopter from about 30 meters above ground level'. Then came the tests with parachutes, some made of paper, some of a biodegradable plastic-like material called Ecofilm. The scientists also tried – in place of parachutes – paper streamers, paper plates, and paper cups.

To track these assorted plummeting agglomerations, the researchers glued a radio transmitter to each mouse's abdomen.

```
┌─ ── ── ── ── ── ── ── ── ── ── ── ── ─┐
│ termine                                │
│ n the forest    Biodegradable Parachutes        │
│ I                   In September-October 2005, 2 aerial bait drops │
│ I               were conducted on 50 x 200 m drop zones; 1 each │
│ I               on AAFB, Tarague Beach Road, and US Naval │
│ idance          Computer and Telecommunications Station Guam │
│ I               (NCTS, Haputo Beach Road). Vegetation on │
│ icent to        NCTS is similar to Tarague. Two types of │
│ e Beach         biodegradable parachutes were evaluated: paper │
│ I·ninsula.      towel, 23.8 x 27.3 cm, A-A-696 Type 1 Singlefold, │
│ I               Lighthouse for the Blind, New Orleans, Louisiana; │
│ I have been     and a plastic-like material, 20.3 x 20.3 cm, │
│ ngan            EcoFilm®, Cortec Corp., St. Paul, Minnesota. Four │
│ I·iant tree on  pieces of cotton thread (3-30.5 cm long and 1-35.6 │
│ I·e sites, 3    cm long) were individually tied to the corners of │
│ In intervals    each parachute type. The threads were knotted and │
└─ ── ── ── ── ── ─ the longer thread was tied to a rear leg of a DNM. ─┘
```

Detail: Methods, with respect to brown tree snakes and toxic deliveries

Each configuration gets the job done, say the team, 'However, a problem with the two parachutes and the paper plate and paper cup is that threads have to be secured to them for attaching the dead mice. This is a time-consuming effort.' The best arrangement for dropping thawed, frozen, poisoned, dead, radio-equipped mice from a helicopter into a tree, they indicate, is simply to attach a paper streamer to some cardboard, and hot-glue the cardboard to a rear leg of the mouse.

Savarie, Peter J., Tom C. Mathies, and Kathleen A. Fagerstone (2007). 'Flotation Materials for Aerial Delivery of Acetaminophen Toxic Baits to Brown Tree Snakes.' *Managing Vertebrate Invasive Species: Proceedings of an International Symposium*, Fort Collins, CO, 7–9 August, 218–23.

SOME CALL IT LOVE, BUT MOST CALL IT NECROPHILIA

John Troyer, a newly arrived scholar at the University of Bath's Centre for Death and Society, dug up evidence of a little-unappreciated gap in the law. His study, called 'Abuse of a Corpse: A Brief History and Re-Theorization of Necrophilia Laws in the USA', appears in the only-occasionally-ghoulish journal *Mortality*.

Troyer spotlights an incident that frustrated the police and the courts of one American state. He writes: 'In September 2006, Wisconsin police discovered Nicholas Grunke, Alexander Grunke, and Dustin Radtke digging into the grave of a recently deceased woman. Upon questioning by police, Alexander Grunke explained that the three men wanted to exhume the body for sexual intercourse. In the Wisconsin state court system, the three men were charged with attempted third-degree sexual assault and attempted theft. None of the men could be charged with attempted necrophilia, since the state of Wisconsin has no law making necrophilia illegal. What the Wisconsin case exposed was the following gap in US jurisprudence: many states have no law prohibiting necrophilia.'

Troyer sketches the court's dilemma: 'Since [the victim] was already dead at the time of the alleged crime and therefore no longer a person before the law, her body was legally recognized as human remains and not as a victim... Nicholas Grunke, Alexander Grunke, and Dustin Radtke were then charged with the remaining illegal acts, namely, damage to cemetery property and attempted theft of movable property, a category that included [the victim's] corpse. What shocked many of the case's observers was that even if the three men had actually succeeded in taking the body from the grave, they could have only been charged with theft of private property since [the victim's] post-mortem body belonged to her parents.'

On 5 March 2008, the Wisconsin Supreme Court heard oral arguments as to how it might overcome the legal lacunae. The session began with a cheery 'We are delighted to welcome with us West Salem High School students'.

The fifty US states differ in their legal grasp of (and on) necrophilia, and the federal government offers them little guidance, but some nations are more organized in their view of the subject. The UK, especially, gives it special focus. Section 70 of the 2003 Sexual Offences Act is entitled 'Sexual Penetration of a Corpse'. Section 70 explicitly, very explicitly, prohibits only the most canonical form of necrophilia. Troyer notes that 'UK law seems to preclude other sexual acts [Troyer lists several of them] from being considered criminal.'

The niceties of the law, especially those pertaining to uncommon activities, suffer a reputation for being abstruse, boring, dull. But Troyer has unearthed an exception. As his study points out: 'Necrophilia is that rare kind of sexual deviancy that truly captures public attention with its abject perversity and titillating, lascivious details.'

Troyer, John (2008). 'Abuse of a Corpse: A Brief History and Re-Theorization of Necrophilia Laws in the USA.' *Mortality* 13 (2): 132–52.

THE AUSTRALIAN FASCINATION WITH CAR CRASHES

Australians are peculiarly fascinated by car crashes, contends Catherine Simpson, a lecturer at Macquarie University in Sydney. Simpson explains how and why in her monograph 'Antipodean Automobility and Crash: Treachery, Trespass and Transformation of the Open Road', which was published in the *Australian Humanities Review*. 'I explore the significance of the car crash in postcolonial Australia', she writes, 'and argue that car accidents are not only presented as an everyday and acceptable form of violence but that the attention to car crashes in Australian films suggests they figure as a moment of rupture in unspoken settler/indigenous violence.'

Australian feature films present hour upon hour of vivid, compelling evidence. Aussie movie crashes explode or unfold in distinctly, proudly Australian ways. The national flair comes across not just in the surrounding scenery but, more important, in the style.

Simpson explains that 'Australia does not have glamorous, Hollywood-style, celebrity car accidents'. She quotes University of Queensland media and cultural studies professor Tom O'Regan on the differences between Australian and US crash cinemachinations: 'Americans dream of freeway pile-ups and their exploitation films have "crazies" driving spectacularly through crowded city streets pursued by slightly crazy policemen ... On the other hand, Australians dream of cars coming over hills in the middle or the wrong side of the road.'

Australia's car crash fascination stems, in part, from its immensity of lonely open space. 'Unlike Europe and many other parts of the world', Simpson says, 'if a vehicle breaks down or crashes in a remote area there is an outside possibility that no one will offer aid ... For most urban-based Australians, the idea of perishing "out there" in the bush after a crash looms much larger than its likelihood ... [This] taps into a deep-seated anxiety in the dominant Australian social imaginary that is connected to the notion of the land as not only hostile but invested with a power to do things to those who venture into it.'

Simpson identifies the 1979 film *Mad Max*, starring Mel Gibson, as one that brought international attention to the Australian car-crash genre. Gibson plays a futuristic-yet-primitive lawman who pursues an evildoer over many miles of road, a relationship that culminates in a spectacular crash and the death of the evildoer; it is enhanced with other car crashes and the deaths of other evildoers, too.

Although not mentioned in the study, Mel Gibson eventually moved to the US, where he had to adapt to American-style cinema car crashes in which, from an Australian's point of view, everyone drives on the wrong side of the road. One could argue – though Simpson does not – that perhaps this intellectual clash of car-crash paradigms led to Mel Gibson's eventual fascination with gory martyrdom, as exhibited in such starring film vehicles as *Braveheart* and *Passion of the Christ*.

Simpson, Catherine (2006). 'Antipodean Automobility and Crash: Treachery, Trespass and Transformation of the Open Road.' *Australian Humanities Review* 39–40.

MAY WE RECOMMEND

'BLOOD AND TISSUE SPATTER ASSOCIATED WITH CHAINSAW DISMEMBERMENT'

by Brad Randall (published in the *Journal of Forensic Sciences*, 2009)

The Royal 'we dismembered two large pig carcasses with a small electric chainsaw in a controlled environment … These experiments have shown that a human body may be easily dismembered with a chainsaw, even a smaller electric-powered model … Despite popular beliefs fueled by crime scene shows on television and recent *Chainsaw Massacre* movies, postmortem dismemberment does not necessarily produce a large amount of blood spatter at a dismemberment scene'.

DEADLINESS OF MONOGRAMS

The great initial discovery – that a man's monogram could cause his early death – was dismaying. But maybe it was all a mistake. A second look, a very careful look, done recently by two skeptical economists, says it just ain't so.

In 1999, the initial blockbuster report, called 'What's in a Name: Mortality and the Power of Symbols', gave certain people the willies. It said: 'individuals with "positive" initials (e.g., A.C.E., V.I.P.) might live longer than those with "negative" initials (e.g., P.I.G., D.I.E.).' Three psychologists at the University of California, San Diego, discovered this by poring through death records, gathering, crunching, and pondering numbers. They then published a warning in the *Journal of Psychosomatic Research*.

Nicholas Christenfeld, David Phillips, and Laura Glynn were deadly serious about it. 'Males with positive initials live 4.48 years longer' than most people, they explained, 'whereas males with negative initials die 2.80 years younger'. They said the effects were smaller for women, perhaps because so many get a change of name when they marry, thus extending their lease on life. None of the research team, it should be noted, has especially interesting full sets of initials themselves.

Christenfeld, Phillips, and Glynn explained the mechanism: 'Parents might fail to notice that the initials they are about to give a child could have negative connotations. This oversight by parents suggests that there may be many offspring who have been inadvertently assigned initials with negative connotations. ... Initials like "A.P.E." or "B.U.M." may cause individuals not to think well of themselves, and the bearers of these initials may have to endure teasing and other negative reactions from those around them.'

Monogrammic Determinism?
STILIAN MORRISON AND GARY SMITH, PHD

Objective: Attempt to replicate a report that people whose names have positive initials (such as ACE or VIP) live much longer than do people with negative initials (such as PIG or DIE). The primary analysis in the original 1969 to 1995 study grouped decedents

Initially, Christenfeld et al. hesitated. 'Upon preliminary inspection', they later wrote, 'the longevity effects appeared too large to be genuine'. But they succumbed, in the end, to the apparently portentous power of their data. (David Phillips, by the way, used some of the same techniques to produce a series of studies in which the data suggested, among other things, that people commonly time their deaths to accord with birthdays or major holidays.)

Gary Smith, an economics professor at Pomona College in Claremont, California, teamed up with his student Stilian Morrison to give the monogram results a good, hard statistical look-see. They looked, they saw, they shook their heads. Then they published their conclusions in the apparently rival journal *Psychosomatic Medicine*.

The Christenfeld study compared the ages of all the people who died in a particular year. But, say Smith and Morrison, if – instead – you look at the lifespans of all the people who were born in a particular year, the pattern doesn't show up. Also, they say, if you use a more complete list of 'good' and 'bad' words (including

Average Age at Death (Sorted from Low to High)
for Each of the Positive and Negative Initials

Average Age at Death	Initials		Number of Cases
	Positive	Negative	
58.90		D.T.H.	194
59.90		S.A.D.	470
62.89		R.A.T.	726
63.11		B.A.D.	539
63.12		D.E.D.	695
63.17		B.U.G.	6
63.33		D.U.D.	3
67.00		U.G.H.	10
67.48		S.I.C.	91
68.09		D.I.E.	45
00.94		M.A.D.	1826
69.02	J.O.Y.		40
69.48		S.I.K.	50
69.76	W.E.L.		659
69.87	W.O.W.		192
69.99		R.O.T.	97
71.36		A.P.E.	94
71.76		P.I.G.	50
72.15		A.S.S.	566
72.62	V.I.P.		97
72.77	H.U.G.		9
72.83	L.I.F.		116
73.06	A.C.E.		376
73.97	G.O.D.		132
74.17	W.I.N.		41
74.67		B.U.M.	12
74.83		I.L.L.	203
75.90	L.I.V.		43
76.03		H.O.G.	123
76.85	L.O.V.		00

T.I.T., G.A.S., variations of the F-word, to name a few) the effect doesn't appear.

If Smith and Morrison are correct in their criticism, the initials big bang theory may have died at the untimely age of six years.

Christenfeld, Nicholas, David P. Phillips, and Laura M. Glynn (1999). 'What's in a Name: Mortality and the Power of Symbols.' *Journal of Psychosomatic Research* 47 (3): 241–54.
Morrison, Stilian, and Gary Smith (2005). 'Monogrammic Determinism?' *Psychosomatic Medicine* 67 (5): 820–24.

THE ECONOMIC ART OF SUICIDE

A study called 'Artists' Suicides as a Public Good' explains how we benefit when a famous artist kills himself. As far as I know, this is the only academic report that credits Kurt Cobain as its major source of information.

Kurt Cobain, the lead singer of the grunge music group Nirvana, committed suicide in 1994 (though, as is traditional when celebrities do themselves in, some people insist it was murder). Professors Samuel Cameron, Bijou Yang, and David Lester theorized about the economic consequences of Cobain's death. Cameron is an economics professor at the University of Bradford, in the UK. Yang and Lester are wife and husband, she an economist at Drexel University in Philadelphia, he a psychologist at the Richard Stockton College of New Jersey.

By almost any numerical measure, Lester is the world's preeminent suicide researcher. Since 1966, he has published more than eight hundred academic reports about suicide. His articles tend to be brief; many are one or two pages long. Typically, Lester analyses statistics that are on the public record. In the 2003 study 'Suicide by Jumping From a Bridge', for example, he reveals that of the 132 suicides jumping from the Delaware Memorial Bridge from 1952 to 2003, the majority were from Delaware.

'Artists' Suicides as a Public Good' is mainly a study in economics, the so-called dismal science, but the paper's tone is almost cheerful. 'The perspective on suicide from the discipline of economics', the report says, 'has to lead us to the position that suicide may be a good thing.'

The three professors walk us through the debits and credits of Kurt Cobain's suicide. Mostly, they see credits: increased sales of his music and associated merchandise; increased 'iconic value' of the products his fans had already purchased; and a variety of emotional benefits which could theoretically be given a financial value. The musicians associated with Kurt Cobain, especially his wife, Courtney Love, and her band, Hole, also presumably benefit from an increase in attention and perceived value.

Lester and his colleagues point out some further, subtle benefits to Kurt Cobain's suicide. Cobain died at twenty-seven, early in a human lifespan, but perhaps fairly late in a pop singer's expected career. 'The potential productivity of his future artistic productivity may be much less than was generated by his suicide', they write. 'Indeed, it is possible that future mediocre works might have blighted a legacy, leading to negative reappraisals and possibly lower sales of his peak-period work.'

Of suicide in general – and specifically of any me-too suicides that Cobain's death may have inspired – the professors describe what they see as a higher sort of economic benefit to society. There is, they write, a 'selective elimination of those who are unable to cope adequately with the requirements of the environment in which they are trying to survive'.

At the end of the study, Lester, Yang, and Cameron mention that they were actually unable to obtain most of the data needed to do their study properly. 'Thus', they write in the report's final sentence, 'at the present time it has been impossible to conduct a methodologically sound study of this phenomenon.'

Cameron, Samuel, Bijou Yang, and David Lester (2005). 'Artists' Suicides as a Public Good.' *Archives of Suicide Research* 9 (4): 389–96.
Lester, David (2003). 'Suicide by Jumping From a Bridge.' *Perceptual and Motor Skills* 97 (1): 338.

MAY WE RECOMMEND

'FATALITIES ATTRIBUTED TO ENTERING MANURE WASTE PITS'

(published in the US National Institute of Health's *Morbidity and Mortality Weekly Report*, 7 May 1993)

GRAVELY MISTAKEN

When a stranger says he wants to dig up a corpse that might be buried beneath the pews of your church, should you let him? Would it help if he explains that: (a) he recently dug up a corpse on the other side of an ocean; and (b) he's not certain who that foreign corpse is, but he thinks it might be a relative of the corpse that might be buried in your church; and (c) he's doing this to bring attention to a man who played an early role in a small, miserable failure four centuries ago?

American historian William M. Kelso thinks you should. Kelso's book *Jamestown – The Buried Truth* tells how (a) he convinced two British churches to let him poke into their bowels; and (b) he also convinced the Church of England to, for the first time in its history, give permission for such poking; and (c) the digging did not proceed smoothly; and (d) the church corpses turned out to be, probably, not the ones he was looking for.

Jamestown, Virginia, was Britain's first real settlement in North America. After a sea voyage filled with flounderings, calamities, and mutinies, the settlers settled in ineptly. Early Jamestown was a catastrophe. Still, in 2007, historians celebrated its four-hundredth anniversary.

Five years earlier, Kelso had dug up an old coffin at Jamestown. The skeleton inside, he speculated, just might be Bartholomew Gosnold, an expedition leader who died a few weeks after arriving. Kelso has a theory that, if only Gosnold had lived, the colony might have succeeded. 'The discovery of his remains', writes Kelso, 'might help inspire a more careful reading of the record of initial English colonization.'

Inspired by this chance to inspire others, Kelso set out to prove that these were Gosnold's bones. His strategy: compare DNA from this inspirational skeleton with DNA from dead Gosnold relatives, if he could find any. 'With skillful deduction from evidence found in various wills and church records', Kelso says, he identified two places he might look. Bartholomew's sister was, possibly, buried under All Saints Church in Shelly, near Ipswich. Bartholomew's niece was, possibly, beneath a church in Stowmarket. Kelso describes at length his battle to obtain all the necessary permissions. 'The international significance of Gosnold won the day', he writes.

Then came the digging. Kelso did find skeletons, but they seem not to be Gosnold's relatives. Of course, Kelso says, he cannot rule out the possibility.

Kelso's book is, mostly, a catalogue of the colony's horrible struggles and difficulties. Its most famous figure, John Smith, was a boastful liar. The colonists, many of them gentlemen ill-prepared to be settlers, achieved failure at almost everything they tried. When the food ran out, they fell to eating 'dogs, cats, rats and mice', and apparently, in the case of at least one husband, his pregnant wife.

And the possibly most important figure, Captain Bartholomew Gosnold, lies buried in mystery, though one cannot with certainty say exactly where.

Kelso, William M. (2006). *Jamestown – The Buried Truth*. Charlottesville: University of Virginia Press.

AN IMPROBABLE INNOVATION

'EDGED NON-HORIZONTAL BURIAL CONTAINERS'
by Donald E. Scruggs (US Patent no. 8,046,883, granted 2011)

Scruggs's previous invention – 'The Easy Inter Burial Container' provided, for the first time, 'a series of burial containers which can be pressed, agitated, screwed and or self bored into a receiving material and provide low cost interment methods' a/k/a a screw-in coffin. The new container makes further progress towards time and space savings.

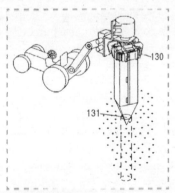

Figure: 'A tractor backhoe using a square section clamping gripper driver to hold, revolve and press a casket into a pre-bored or augered hole

HE ATE THE SILVERWARE

A study called 'Account of a Man Who Lived Ten Years After Having Swallowed a Number of Clasp-Knives, with a Description of the Appearances of the Body after Death', published in 1823, has an accurate title. But in a sense, it is misnamed. The author could, with just as much accuracy, have chosen to call it 'Account of a Man Who Died Ten Years After Having Swallowed a Number of Clasp-Knives'.

The author, Alex. Marcet, MD, FRS, etc., was a London physician. This monograph is his most enduring legacy to scholars, doctors, and perhaps also to pocket-knife manufacturers. Marcet called it 'a most striking illustration of the self-preserving powers of the stomach and intestines'.

The hero – or at least the central figure – of the story is John Cummings, an American sailor who died in 1809 in Guy's Hospital under the care of Marcet's colleague Dr Curry. Marcet wrote about what had happened. He based his account partly on accounts by Curry, and partly on 'a narrative, written with great distinctness and simplicity, by the patient himself' that was 'found in the patient's pocket after his decease'.

One day, on shore leave in France with some buddies, Cummings saw an entertainer pretend to swallow clasp knives. Later, 'after drinking freely', Cummings 'boasted that he could swallow knives as well as the Frenchman'. With goading from his companions, Cummings swallowed his own pocket knife. After further encouragement, he ate three more.

Three of the knives emerged fairly soon from his digestive system. The fourth did not.

Six years later, at the behest of crowds of revellers, Cummings swallowed fourteen clasp knives in Boston, Massachusetts. In subsequent weeks, Cummings was, as he himself put it, 'safely delivered of his cargo'.

The basic pattern repeated many times. Knives entered the system. Some left. Some stayed.

Time, tides, and fair winds later brought Cummings to England. Several times he was admitted for treatment at Guy's Hospital. In 1808 he 'became a patient of Dr Curry, under whose care he remained, gradually and miserably sinking under his sufferings, till March 1809, when he died, in a state of extreme emaciation'.

Marcet's report includes a handsome drawing that shows thirty-eight objects, some clearly identifiable as knife parts, retrieved during an autopsy of the late Mr Cummings.

The Cummings cutlery collection, notable at the time, now seems almost paltry when compared with achievements in later centuries. Most impressive, perhaps, is the collection of seventy-eight forks and spoons (but no knives) removed from the innards of someone who also swallowed salt-and-pepper-shaker tops and more than a thousand additional items. These and other rare collectibles are on display at the Glore Psychiatric Museum in St Joseph, Missouri. They are a silver-plated example of the excesses of consumer mentality.

I should thank Dan Meyer, president of the Sword Swallowers Association International and Ig Nobel Prize-winning co-author of the *BMJ* study 'Sword Swallowing and Its Side Effects', for demonstrating some of the fine points of this research.

Marcet, Alex. (1823). 'Account of a Man Who Lived Ten Years After Having Swallowed a Number of Clasp-Knives, with a Description of the Appearances of the Body after Death.' *Medico-Chirurgical Transactions* 12 (1): 52–63.

Witcombe, Brian, and Dan Meyer (2006). 'Sword Swallowing and Its Side Effects.' *BMJ* 333. 1285–87.

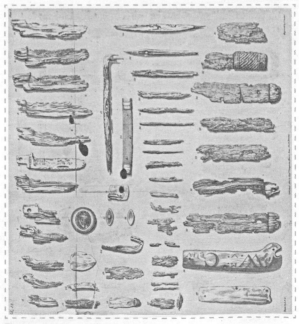

The eponymous number of clasp knives

JUST ADD. WHAT, ER?

IN BRIEF

'A CUTE CHARACTERIZATION OF ACUTE TRIANGLES'
by H. J. Seiffert (published in the *American Mathematical Monthly*, 2000)

Some of what's in this chapter: Dry, laundry, dry • The rise and fall of hairdresser-mathematicians • Ham sandwiches in the minds of mathematicians • How to pour another cup of coffee • Lead versus feathers • Statistical faces • Beautiful maths in the mouth • Cheddar cheese, mechanically • A tittle obliquity measurer • The lazy bureacrat problem • Promoting people, randomly

ON THE DRYING OF LAUNDRY

'It is striking that the drying process familiar to most people, namely, that of drying laundry hung from a clothesline, does not seem to have been investigated in a quantitative, scientific manner.'

With those words, and many more, Eric B. Hansen introduced a generation to the subtle mathematical pleasures of damp cloth. Hansen's treatise 'On Drying of Laundry' holds a treasured place in the collections and hearts of countless mathematicians, engineers, and persons of the cloth. Countless, certainly, for there is no reliable way to count, even very roughly, how many individuals own copies (or copies of copies) of the report, and how many have read borrowed copies, and how many others have learned its contents by word-of-mouth.

The ten-page paper presumably delighted the readers of the *SIAM Journal on Applied Mathematics* when it appeared in the October 1992 issue. But to some of them, it must have been a matter of anticipation being satisfied. Eric Hansen had discussed the drying of laundry a full two years earlier, in Montreal, at the International Conference on Free Boundary Problems. So far as I have determined, that was the first time that drying laundry had been aired in public, at least in a mathematically correct manner.

Hansen was based at the Technical University of Denmark, and his work must have been the talk of the event, despite the inclusion there of other, even more dry, presentations. That was the year of Klarbring, Mikelic, and Shillor's 'On the Rigid Punch with Friction', and M. Chipot's 'New Remarks on the Dam Problem'. (Chipot gave utterance in a plenary session. You can learn details by reading a copy of the *Proceedings of the Montreal Meeting on Free Boundary Problems*.)

Drying laundry is a complex and subtle phenomenon. Hansen did a laudable job of keeping it clear and relatively simple. Equations are of course clearer and more concise than prose, and Hansen managed to transform what would have been a huge amount of dense prose, wringing it down into twenty-one crisp, clean equations.

Skeptics may try to write this off as a 'mere' theoretical exercise, but they would be wrong. Hansen went way beyond theory. He performed an experiment with a wet T-shirt. He reports that the results agree well with his theoretical predictions.

The examination and analysis of wet T-shirts is something that many non-scientists believe they understand, at least on a practical level. But Eric Hansen's delightful paper suggests that when a scientist looks at a wet T-shirt, he deeply appreciates it.

Hansen, Erik B. (1992). 'On Drying of Laundry.' *SIAM Journal on Applied Mathematics* 52 (5): 1360–69.

Chipot, M. (1993). 'New Remarks on the Dam Problem.' In Chadam, John M., and Henning Rasmussen, eds. *Emerging Applications in Free Boundary Problems: Proceedings of the International Colloquium 'Free Boundary Problems: Theory and Applications.'* Pitman Research Notes in Mathematics Series 280. Harlow, UK: Longman Scientific and Technical: 2–12.

Klarbring, A., A. Mikelic, and M. Shillor (1993). 'On the Rigid Punch with Friction.' In Chadam, John M., and Henning Rasmussen, eds. *Emerging Applications in Free Boundary Problems: Proceedings of the International Colloquium 'Free Boundary Problems: Theory and Applications.'* Pitman Research Notes in Mathematics Series 280. Harlow, UK: Longman Scientific and Technical: 35–40.

A BRUSH WITH MATHS

Some mathematicians pay attention to hairdressers more than other mathematicians do. Two modern scholars focused their attention very differently when they wrote about history's most famous numerico-tonsorial collaboration.

In 1784, mathematicians joined forces with hairdressers on a scale probably never attempted before or since. A century and a half later, Raymond Clare Archibald looked back at their forces in wonder. Archibald's 'Tables of Trigonometric Functions in Non-Sexagesimal Arguments' spanned twelve full pages in the April 1943 issue of the about-as-lively-as-you-might-expect journal *Mathematical Tables and Other Aids to Computation*.

Archibald, a professor of mathematics at Brown University in Providence, Rhode Island, exemplified terseness. He often abbreviated himself as R. C. Archibald. This monograph identifies him simply as 'RCA'.

Though some mathematicians are bald, he was not. A former student wrote that Archibald 'was striking in appearance, his hair wavy and beginning to grey, worn a little longer than was generally the custom'.

Archibald sketched the basics of the hairdresser story.

The French government wanted new, improved 'tables of the sines, tangents, etc., and their logarithms'. The fellow in charge, the not-so-terse Gaspard Clair François Marie Riche de Prony, whose portraits credit his head with an abundant, curvilinear garden of hair, assembled a team. De Prony got three or four mathematicians to do the heavy mental lifting, seven or eight people to perform the tedious calculations, and – here the story took its little twist – '70 or 80' people to check the work.

These checkers, Archibald said, were 'endowed with no great mathematical abilities. In fact they were mainly recruited from among hairdressers whom the abandonment of the wig and powdered hair in men's fashions, had deprived of a livelihood.'

Archibald devoted only one paragraph to those hairdressers, otherwise stubbornly persisting in an almost obsessive description of the sines, cosines, and other, sometimes tangential, niceties of the story. The project produced '17 large folio volumes', he has us know, of which '8 volumes were devoted to logarithms of numbers to 200,000'.

In contrast, Ivor Grattan-Guinness practically babbled about the ex-coiffeurists. An emeritus professor of history of mathematics and logic at Middlesex Polytechnic, Grattan-Guinness sports healthy hanks of white hair in the photos of him that I have seen. His monograph called 'Work for the Hairdressers: The Production of de Prony's Logarithmic and Trigonometric Tables' appeared in 1990 in the *Annals of the History of Computation*. He wrote: 'Many of these workers were unemployed hairdressers: one of the most hated symbols of the *ancien regime* was the hairstyles of the aristocracy, and the obligatory reduction of coiffure "as the geometers say, to its most simplest expression" left the hairdressing trade in a severe state of recession. Thus these artists were converted into elementary arithmeticians.'

Everything was carefully organized, Grattan-Guinness explained, 'to avoid multiplication and division and to reduce the calculations to sums and (especially) differences, which the hairdressers could fairly be expected to handle'.

The hairdressers finished their work in less than three years. Historians have (so far as I'm aware) ignored whatever they did after that.

Archibald, Raymond Clare (1943). 'Tables of Trigonometric Functions in Non-Sexagesimal Arguments.' *Mathematical Tables and Other Aids to Computation* 1 (2): 33–44.
Grattan-Guinness, Ivor (1990). 'Work for the Hairdressers: The Production of de Prony's Logarithmic and Trigonometric Tables.' *Annals of the History of Computation* 12: 177–85.

LEFTOVERS FROM HAM SANDWICH THEORIES

The Ham Sandwich Theorem has been a treat and a spur to mathematicians for more than half a century. There was a bit of a kerfuffle about who invented it, but that question did get settled.

The Ham Sandwich Theorem cropped up in a branch of mathematics called algebraic topology.

The theorem describes a particular truth about certain shapes. Most published papers on the topic make a hash of explaining it to anyone who is not an algebraic topologist. But the authors of a 2001 paper called 'Leftovers from the Ham Sandwich Theorem' wrapped up an important little leftover – they put the idea into clear language.

The Ham Sandwich Theorem, they wrote, 'rescues the careless sandwich maker by guaranteeing that it is always possible to slice the sandwich with one cut so that the ham and both slices of bread are each divided into equal halves, no matter how haphazardly the ingredients are arranged'.

For a while, most ham sandwich theorizing dealt with simple cases. A paper entitled 'Computing a Ham-Sandwich Cut in Two Dimensions', published in 1986 in the *Journal of Symbolic Computation,* is typical. It considered only ham sandwiches that had been flattened flatter than even the chintziest cook would dare to devise. Mathematicians often do things this way, first considering the extreme cases, digesting those thoroughly, and only then moving on to more substantial versions. Indeed, the 'Computing a Ham-Sandwich Cut in Two Dimensions' paper itself contains a section called 'Getting Rid of Degenerate Cases'.

People did solve the mystery of slicing a thick ham sandwich. And, inevitably, they developed a hunger for more substantial problems.

In 1990, Yugoslavian theorists were writing in the *Bulletin of the London Mathematical Society* about 'An Extension of the Ham Sandwich Theorem'. Two years later a theorist at Yaroslav State University in Russia published a paper called 'A Generalization of the Ham Sandwich Theorem'. That same year, a team of hungry American, Czech, and German mathematicians assembled a master collection of recipes for slicing ham sandwiches. Mathematicians almost never use the word 'recipe', so they called their paper 'Algorithms for Ham-Sandwich Cuts'. You'll find it in the December 1994 issue of the journal *Discrete and Computational Geometry.*

Research then moved on to exotic, distantly related questions, exemplified by a 1998 monograph called 'Green Eggs and Ham'.

And who started this? A 2004 paper called 'The Early History of the Ham Sandwich Theorem' took care of a lingering leftover: it

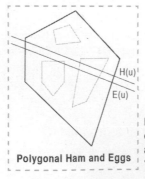

Polygonal Ham and Eggs

Figure: 'The knife cut which divides the ham into equal area portions' as depicted in 'Green Eggs and Ham'

identified the inventor. Mathematico-historians W. A. Beyer and Andrew Zardecki, of Los Alamos National Laboratory in New Mexico, say that it was a Jewish theorist who introduced the ham sandwich into mathematical theory. Beyer and Zardecki trace the theorem back to a 1945 paper by the Polish mathematician Hugo Steinhaus that 'represents work Steinhaus did in Poland on the ham sandwich problem in World War II while hiding out with a Polish farm family'.

Byrnes, Graham, Grant Cairns, and Barry Jessup (2001). 'Leftovers from the Ham Sandwich Theorem.' *American Mathematical Monthly* 108 (3): 246–49.

Beyer, W. A., and Andrew Zardecki (2004). 'The Early History of the Ham Sandwich Theorem.' *American Mathematical Monthly* 111 (1): 58–61.

Edelsbrunner H., and R. Waupotitsch (1986). 'Computing a Ham-Sandwich Cut in Two Dimensions.' *Journal of Symbolic Computation* 2 (2): 171–78.

Zivaljevic, Rade T., and Sinisa T. Vrecica (1990). 'An Extension of the Ham Sandwich Theorem.' *Bulletin of the London Mathematical Society* 22 (2): 183–86.

Dolnikov, V. L., and P. G. Demidov (1992). 'A Generalization of the Ham Sandwich Theorem.' *Matematicheskie Zametki* 52 (2): 27–37.

Lo, Chi-Yuan, J. Matoušek, and W. Steiger (1994). 'Algorithms for Ham-Sandwich Cuts.' *Discrete and Computational Geometry* 11 (1): 433–52.

Kaiser, M. J., and S. Hossaien Cheraghi (1998). 'Green Eggs and Ham.' *Mathematical and Computer Modeling* 28 (1): 91–99.

Abbott, Timothy G., Michael A. Burr, Timothy M. Chan, Erik D. Demaine, Martin L. Demaine, John Hugg, Daniel Kane, Stefan Langerman, Jelani Nelson, Eynat Rafalin, Kathryn Seyboth, and Vincent Yeung. 'Dynamic Ham-Sandwich Cuts in the Plane.' *Computational Geometry* 42 (5): 419–28.

Steiger, William, and Jihui Zhao (2009). 'Generalized Ham-Sandwich Cuts.' *Discrete and Computational Geometry*. 44 (3): 535–45.

MAY WE RECOMMEND

GREEK RURAL POSTMEN AND THEIR CANCELLATION NUMBERS
edited by Derek Willan (publication of the Hellenic Philatelic Society of Great Britain, 1994)

THE PERFECT SECOND CUP OF COFFEE

Yes, there is a best way – mathematically – to pour your second cup of coffee, says a study called 'Recursive Binary Sequences of Differences'.

But no one realized it until the year 2001, when Robert M. Richman published his simple recipe in the journal *Complex Systems*. During the subsequent passage of nine years and billions of cups of coffee, the secret has been available to all.

'The problem is that the coffee that initially comes through the filter is much stronger than that which comes out last, so the coffee at the bottom of the pot is stronger than that at the top', says Richman. 'Swirling the pot does not homogenize the coffee, but using the proper pouring pattern does.'

Here's all you have to do. Prepare coffee – two cups' worth – in a carafe. Now get two mugs, call them A and B. Then: 'If one has the patience to make four pours of equal volume, the possible pouring sequences are AABB, ABBA, and ABAB.'

Choose ABBA.

That's it. You now have two nearly-identical-tasting cups of coffee.

Richman tells you what to do if you're pernickety: 'If one wishes to further reduce the difference and has more patience, one can make eight pours of equal volume, four in each cup. The number of possible sequences is now 35.' The optimal sequence, he calculates, is ABBABAAB.

And if you are more finicky than that, Richman neglects you not. 'With even more patience, one may make 16 pours, eight into each cup. There are now 6435 possible pouring sequences.' ABBABAAB-BAABABBA is the way to go.

This same blending problem crops up elsewhere in modern life: in distributing pigments evenly when mixing paint, and even in choosing sides for a basketball game. 'Consider the fairest way for "captain A" and "captain B" to choose sides,' Richman instructs. The traditional method – alternating the choices – leads to unequally strong teams. Instead use the coffee recipe, which is 'likely to result in the most equitable distribution of talent'. Insist that 'captain A has the first, fourth, sixth, and seventh choices, while captain B has the second, third, fifth, and eighth choices'.

The mathematics in this study looks at coffee production as a collection of 'Walsh functions'. These are trains of on/off pulses that add together in enlightening ways.

The monograph ends modestly, or perhaps realistically, with a wistful thought: 'As is typically the case with fundamental contributions, scientifically significant applications of this work may not appear for some time.'

Richman recently retired as a chemistry professor at Mount St. Mary's University in Emmitsburg, Maryland. He now has more time to devote to this mixing business, with pleasure. 'It took me over ten years to develop the mathematics to solve this problem, which is well outside of my primary area of expertise. I'm trying to find a classical number theorist who is willing to collaborate on the sequel: I think I can definitively establish the best way to pour three cups of coffee.'

Richman, Robert M. (2001). 'Recursive Binary Sequences of Differences.' *Complex Systems* 13: 381–92.

THE FULL WEIGHT OF SCIENCE

A pound of lead feels heavier than a pound of feathers – a thing long suspected, but not carefully tested until 2007, when Jeffrey B. Wagman, Corinne Zimmerman, and Christopher Sorric ran an experiment involving lead, feathers, plastic bags, cardboard boxes, a chair, blackened goggles, and twenty-three volunteers from the city of Normal, Illinois.

The scientists are based at Illinois State University, which is located in that unassumingly named metropolis. In a study published in the journal *Perception*, they explain why they took the trouble. '"Which weighs more – a pound of lead or a pound of feathers?" The seemingly naïve answer to this familiar riddle is the pound of lead whereas the correct answer is that they weigh the same amount.' But, they wrote, this 'naïve answer may not be so naïve after all. For over 100 years, psychologists have known that two objects of equal mass can feel unequally heavy depending on the mass distribution of those objects.'

Wagman, Zimmerman, and Sorric poured some lead shot into a plastic bag, then sealed and taped the bag inside the bottom of a cardboard box. For clarity, let's call this the box-with-lead-in-its-bottom. Then they stuffed a pound of goose down feathers into a large plastic bag. Feathers and bags being what they are, this fluffed, baggy entity entirely filled a box that looked just like the box-with-lead-in-its-bottom. Let's call this snugly packed second box the box-with-feathers-spread-thoughout-its-innards.

Then came the test. One by one the volunteers sat in the chair, donned the blackened goggles, then 'placed the palm of their preferred hand up with their fingers relaxed. On a given trial, each box was placed on the participant's palm in succession. The participant hefted each box and reported which box felt heavier.'

Slightly more often than not, the volunteers said that the box-with-lead-in-its-bottom was heavier than the box-with-feathers-spread-thoughout-its-innards.

After weighing and judging all the data, the scientists educatedly hazarded a guess as to why one box seemed heavier. Probably, they said, it's because 'the mass of the feathers was distributed more or less symmetrically in the box (i.e., the feathers filled the box), but the mass of the lead was distributed asymmetrically along the vertical axis (i.e., the box was "bottom-heavy"). Therefore the box containing lead was more difficult to control, and it felt heavier.'

The scientists did not test how volunteers would respond if the lead were fixed precisely in the middle, rather than stuck to the bottom, of the box. This they left for future scientists to contemplate.

Wagman, Jeffrey B., Corinne Zimmerman, and Christopher Sorric (2007). '"Which Feels Heavier – A Pound of Lead or a Pound of Feathers?" A Potential Perceptual Basis of a Cognitive Riddle.' *Perception* 36: 1709–11.

MAY WE RECOMMEND

'DO DOGS KNOW CALCULUS?'
by Timothy J. Pennings (published in *College Mathematics Journal*, 2003) and

'DOGS DON'T NEED CALCULUS'
by Michael Bolt and Daniel C. Isaksen (published in *College Mathematics Journal*, 2010)

THE FACE VALUE OF NUMBERS

A smiley-face is very expressive, statistically. By tweaking the eyes, mouth, and other bits, you can literally put a meaningful face on any jumble of numbers. Herman Chernoff pointed this out in 1973 in the *Journal of the American Statistical Association*, in an article entitled 'The Use of Faces to Represent Points in K-Dimensional Space Graphically'.

Subsequently, folks took to calling these things Chernoff faces. Chernoff faces can make statistical analysis into a recognizably human activity.

Most people, when shown some statistics, sigh and get boggled. But Herman Chernoff realized that almost everyone is good at reading faces. So he devised recipes to convert any set of statistics into an equivalent bunch of smiley-face drawings.

Each data point, he wrote, 'is represented by a cartoon of a face whose features, such as length of nose and curvature of mouth, correspond to components of the point. Thus every multivariate observation is visualized as a computer-drawn face. This presentation makes it easy for the human mind to grasp many of the essential regularities and irregularities present in the data.'

'The Use of Faces to Represent Points in K-Dimensional Space Graphically' is one of the few statistics papers that is visually goofy, rather than arid.

One page is filled with eighty-seven cartoon faces, each slightly different. Some faces have little beady eyes, others have big, startled, wide-awake peepers. There are wide mouths, little dried-up 'I'm not here, don't notice me' mouths, and middling mouths. Another page shows off some of the cartoony variety that's possible: roundish simpleton heads, jowly alien-visitor heads, and a smattering of noggins that look froggy. Elsewhere, the study perhaps inevitably includes conventional statistics machinery – charts of numbers, differential and integral calculus equations, and plenty of technical lingo.

Chernoff discovered, by experiment, that people could comfortably interpret a face that expresses quite large amounts of data. 'At this point,' he wrote, 'one can treat up to eighteen variables, but it would be relatively easy to increase that number by adding other features such as ears, hair, [and] facial lines.'

The world has gone on to employ Chernoff faces a little, but not yet a lot. A 1981 report in the *Journal of Marketing*, for example, used them to display corporate financial data, with this explanation: 'From Year 5 to Year 1, the nose narrows as well as increases in length, and the eccentricity of the eyes increases. Respectively, these facial features represent a decrease in total assets, an increase in the ratio of retained earnings to total assets, and an increase in cash flow.'

**Facial Representation of Financial Performance
(1 to 5 Years Prior to Failure)**

Federal US

Date Dimensions	Year to Failure				
	5	4	3	2	1
1. Return on Assets	0.10	0.11	0.06	0.03	-0.16
2. Debt Service	3.66	3.79	1.55	0.78	-14.11
3. Cash Flows	1.53	1.48	1.39	1.35	0.94
4. Capitalization	0.22	0.20	0.18	0.16	-0.02
5. Current Ratio	71.40	89.10	97.85	96.80	58.21
6. Cash Turnover	24.03	25.92	25.62	27.40	71.26
7. Receivables Turnover	5.25	4.46	4.26	4.36	9.56
8. Inventory Turnover	5.38	4.77	4.57	4.44	5.34
9. Sales per Dollar Working Capital	6.74	6.33	7.02	7.61	-45.77
10. Retained Earning/Total Assets	0.32	0.30	0.01	-0.01	-0.26
11. Total Assets	0.94	0.76	0.39	0.45	0.43

A note at the very end of Chernoff's 1973 paper hints at a practical reason why his idea would not catch on immediately: 'At this time the cost of drawing these faces is about 20 to 25 cents per face on the IBM 360-67 at Stanford University using the Calcomp Plotter. Most of this cost is in the computing, and I believe that it should be possible to reduce it considerably.'

Chernoff, Herman (1973). 'The Use of Faces to Represent Points in K-Dimensional Space Graphically.' *Journal of the American Statistical Association* 68 (342): 361–68.
Huff, David L., Vijay Mahajan, and William C. Black (1981). 'Facial Representation of Multivariate Data.' *Journal of Marketing* 45 (4): 53–59.

A GOLDEN MEAN IN YOUR MOUTH

Eddy Levin of Harley Street in London puts a golden ratio, not just golden teeth, into people's mouths. Dr Levin has been at this for a while. It was he who wrote a study called 'Dental Esthetics and the Golden Proportion', which graced pages 244–252 of the September 1978 issue of the *Journal of Prosthetic Dentistry*.

The golden ratio is a special number that has caught the eye and imagination of mathematicians, artists, and now, thanks to Levin, dentists. Some call it the 'golden mean' (though philosophers use that phrase to mean something else); some call it the 'golden section'. Some Germans call it, evocatively, the *goldener Schnitt*. Most everyone calls it beautiful.

The golden ratio is the number you get when you compare the lengths of certain parts of certain perfectly beautiful things (among them: snail shell spirals, the Parthenon in Athens, and Da Vinci's 'The Last Supper'). You'll find that the ratio of the bigger part to the smaller equals the ratio of the combined length to the bigger. That ratio, that number, is always the same, ever so slightly bigger than 1.6180339.

If doing sums causes you pain, just go find someone who has perfect teeth and who won't mind you staring into his or her mouth.

Levin explains that many years ago he was both studying maths and trying to find out what made teeth look beautiful. 'It was at a moment,' he says, 'like when Archimedes got into his bath, that I suddenly realized that the two were connected – the Golden Proportion and the beauty of teeth. I began to put this into practise and started testing my ideas on my patients. My first case was a young girl in a hospital, where I was teaching, whose front teeth were in a terrible state and needed crowning. Despite the scepticism of the other members of staff and the unenthusiastic technicians with whom I had to work and whose co-operation I depended upon, I crowned all her front teeth, using the principles of the Golden Proportion. Everybody, including the young lady herself, agreed that her teeth now looked magnificent.'

Most important, in Levin's reckoning, is the simple tooth-to-tooth ratio: 'The four front teeth, from central incisor to premolar are the most significant part of the smile and they are in the Golden Proportion to each other.'

Levin created an instrument called the 'golden mean gauge'. Made of stainless steel 1.5 millimetres thick, and retailing for £85 (about $135), it shows whether the numerous major dental landmarks 'are in the Golden Proportion', and it is suitable for autoclaving. He also offers a larger version that is 'useful for full face measurements' and 'useful to measure larger objects or bigger pictures or furniture etc.'

Levin, E. I. (1978). 'Dental Esthetics and the Golden Proportion.' *Journal of Prosthetic Dentistry* 40 (3): 244–52.

IN BRIEF

'DISCOVERING INTERESTING HOLES IN DATA'

by Bing Liu, Liang-Ping Ku, and Wynne Hsu (*Proceedings of Fifteenth International Joint Conference on Artificial Intelligence*, Nagoya, Japan, 1997)

The authors, at the National University of Singapore, explain: 'Clearly, not every hole is interesting ... However, in some situations, empty regions do carry important information'.

CHEESE STRING THEORY

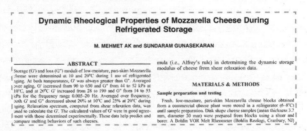

Dynamic Rheological Properties of Mozzarella Cheese During Refrigerated Storage

M. MEHMET AK and SUNDARAM GUNASEKARAN

——————— ABSTRACT ———————

Storage (G') and loss (G") moduli of low-moisture, part-skim Mozzarella cheese were determined at 10 and 20°C during 1 mo of refrigerated aging. At both temperatures, G' was always greater than G". Averaged over aging, G' increased from 90 to 630 and G" from 44 to 52 kPa at 10°C, and at 20°C G' increased from 28 to 190 and G" from 14 to 53 kPa for the frequency range 0.005–20 Hz. Averaged over frequency, both G' and G" decreased about 20% at 10°C and 25% at 20°C during aging. Relaxation spectrum, computed from shear relaxation data, was used to calculate the G'. The calculated values of G' were in good agreement with those determined experimentally. These data help predict and compare melting behaviors of such cheeses.

mula (i.e., Alfrey's rule) in determining the dynamic storage modulus of cheese from shear relaxation data.

MATERIALS & METHODS

Sample preparation and testing

Fresh, low-moisture, part-skim Mozzarella cheese blocks obtained from a commercial cheese plant were stored in a refrigerator (6–8°C) until sample preparation. Disk shape cheese samples (mean thickness 3.2 mm, diameter 50 mm) were prepared from blocks using a slicer and borer. A Bohlin VOR Melt Rheometer (Bohlin Reologi, Cranbury, NJ)

January 1995 was a signal month for the understanding of cheese. Maria N. Charalambides and two colleagues, J. G. Williams and S. Chakrabarti, published their master work: 'A Study of the Influence of Ageing on the Mechanical Properties of Cheddar Cheese'. It showed a refined way to do mathematical calculations about cheese.

Charalambides is a senior lecturer in the department of mechanical engineering at Imperial College, London. Her report begins with a two-page review of certain incisive cheese studies of the past. The aim of those studies, generally, was to compress a hunk of cheese between two plates, to see what the cheese would do.

This is painstakingly technical work. In 1976, researchers named Culioli and Sherman 'reported a change in the stress-strain behaviour of Gouda cheese when plates were lubricated with oil as opposed to when they were covered with emery paper'. Two years later, Sherman and a different collaborator did similar work with Leicester cheese. Subsequently, other scientists performed related experiments on mozzarella cheese, cheddar cheese, and processed cheese.

The plates and the cheese rub and stick against each other. Their fric tion leads the cheese to warp – to bow outwards or flex inwards – when it's under pressure. And this warp drives scientists half-mad. Frictionless cheese would be easier to study ... but frictionless cheese does not exist.

'It is obvious,' Charalambides writes, 'that quantifying frictional effects in compression tests of cheese is a complicated matter.' Compli- cated, yes – but Charalambides et al. managed to do it.

They compressed cheese cylinders of various heights, calculated the stresses and strains in each of them, and then plotted a mathematical family of cheese stress-strain curves. Some further, almost mundane, calculations yielded up a delicious holy grail of cheese data: a way to es- timate how cheese, minus the effects of friction, behaves under pressure.

Then came the main event: measuring how cheese behaviour changes as the cheese goes from infancy to old age. It would be a happy cheese manufacturer who could reliably gauge a cheese's age by doing a simple mechanical test.

Charalambides and her team performed fracture tests on the cheese, too. Those and the compression tests, done on cheeses young and old, produced a numerical portrait of cheese behaviour from birth through the ripe age of seven months.

The Charalambides report is a deeply pleasurable read for anyone who lives and breathes cheese and has a modest working knowledge of materials science. But those who care deeply about their cheese noted that the study looked at merely three varieties: mild cheddar, sharp cheddar, and Monterey jack.

The next year, mozzarella enthusiasts must have scrambled to buy copies of the May/June 1996 issue of the *Journal of Food Science*, where they could read M. Mehmet Ak and Sundaram Gunasekaran's, 'Dynamic Rhe- ological Properties of Mozzarella Cheese During Refrigerated Storage'.

Since then many scientists have compressed and fractured many kinds of cheese, delving even into the realm of soft cheeses. Mathemat- ics-based mechanical cheese testing is no longer just a romantic dream.

Charalambides, M. N., J. G. Williams, and S. Chakrabarti (1995). 'A Study of the Influence of Ageing on the Mechanical Properties of Cheddar Cheese.' *Journal of Materials Science* 30. 3959–67.
Ak, M. Mehmet, and Sundaram Gunasekaran (1996). 'Dynamic Rheological Properties of Moz- zarella Cheese During Refrigerated Storage.' *Journal of Food Science* 61 (3): 566–69.
Lee, Siew Kim, Skelte Anema, and Henning Klostermeyer (2004). 'The Influence of Moisture Content on the Rheological Properties of Processed Cheese Spreads.' *International Journal of Food Science & Technology* 39 (7): 763–71.

AN IMPROBABLE INNOVATION

'A TITTLE OBLIQUITY MEASURER'

by Zhengcai Li (International patent application no. PCT/
CN2007/003282, filed 2007)

Li, from Tianjin, China, dotted all his i's in the filing: 'When the tittle
obliquity measurer is tilted, the gravity pendulum rotates all the gears,
and the bottom surface of the housing may be normal to the central
line of a vertical shaft which is to be measured, the bottom surface
of the housing may be normal to and crossed with the approximation
plumb surface at the horizontal line which is parallel to the power
shaft, the indication device can indicate the obliquity.'

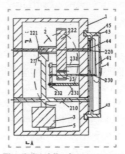

The tittle obliquity measurer

MEASURING UP RULERS

Complimentary small plastic rulers, being imprecise, innacurate,
flimsy, and defaced with advertising, draw only a measured amount
of respect from metrologists. In 1994, two metrologists took measures
to see exactly how much respect the rulers deserve.

Metrologists are the people who come up with more accurate,
more precise ways to measure things.

The metrology community incessantly tussles about new standards definitions for the intimidatingly important, never-quite-as-good-
as-they-ideally-could-be standards – most famously, the kilogram, the
metre, and the second.

The father-and-son team of T. D. Doiron and D. T. Doiron looked,
briefly, at a neglected standard. Their report, called 'Length Metrology

of Complimentary Small Plastic Rulers', drew some measure of interest when it was presented at the Measurement Science Conference in Pasadena, California, in 1994.

Theodore Doiron was a member of the Dimensional Metrology Group at the US National Institute of Standards and Technology. Daniel was, at the time, a teenager at school.

The Doiron/Doiron report implies two simultaneous and opposite truths. Metrologists sometimes express contempt for small plastic rulers (known in the trade as SPRs), because they are made of cheap polystyrene and manufactured to loose tolerances. But metrologists also, down deep, harbour respect for these stylish, useful, slim, flat-bottomed objects, with the four straight-edge working surfaces and the top that boasts a sufficiency of both inked markings and raised graduations, said gradations being located at the outer edges of the bevelled top sides.

The Doirons explain this ambivalent attitude: 'There are virtually no active scientists or engineers who do not have a number of SPRs in their desks which are used continually for developing the earliest and most basic designs of virtually every object manufactured. A quick survey of engineers will show that these early sketches, the very basis of our manufacturing economy, are largely dependent on the use of SPRs. While there is a national standard for plastic rulers, Federal Specification GG-R-001200-1967 and the newer A-A-563 (1981), there has never been a systematic study of the metrology of this basic tool of the national measurement system.'

Doiron and Doiron studied fifty rulers they had 'collected over a long period of time at conferences and from colleagues'. They discovered that the government specification was itself so shockingly poor that they could point to a key passage and say: 'We cannot figure out what this statement means.'

After measuring things as best they could (and being good metrologists, they could measure things well indeed) the Doirons reached a pair of conclusions. First that most of the complimentary small plastic rulers 'quite easily' met the official (albeit murky) standard. Second, that 'the older the ruler' was, the more accurate it was likely to be.

The National Institute of Standards and Technology (NIST) itself, an official told me, once ordered a batch of complimentary small plastic rulers that turned out, upon arrival, to be wretchedly calibrated. As a measure of caution (they have a reputation to protect), and perhaps with some umbrage and embarrassment, NIST returned them to the manufacturer.

Doiron, Daniel T., and Theodore D. Doiron (1994). 'Length Metrology of Complimentary Small Plastic Rulers.' *Proceedings of the Measurement Science Conference*, Anaheim, Calif.

THE LAZY BUREAUCRAT PROBLEM

The lazy bureaucrat problem is ancient, as old as bureaucracy itself. In the 1990s, mathematicians decided to look at the problem. They have since made progress that, depending on your point of view, is either impressive or irrelevant.

Four scientists at the State University of New York, Stony Brook, issued the first formal report. 'The Lazy Bureaucrat Scheduling Problem', by Esther Arkin, Michael Bender, Joseph Mitchell, and Steven Skiena, appeared in the journal *Algorithms and Data Structures*. The study describes a prototypically lazy bureaucrat, transforming this annoying person into a collection of mathematical formulas, theorems, proofs, and algorithms.

jobs. For the LBP we consider three different objective functions, which naturally arise from the bureaucrat's goal of inefficiency:

1. *Minimize the total amount of time spent working* — This objective naturally appeals to a "lazy" bureaucrat.

2. *Minimize the weighted sum of completed jobs* — In this paper we usually assume that the weight of job i is its length, t_i; however, other weights (e.g., unit weights) are also of interest. This objective appeals to a "spiteful" bureaucrat whose goal is to minimize the fees that the company collects on the basis of his labors, assuming that the fee (in proportion to the task length, or a fixed fee per task) is collected only for those tasks that are actually completed.

'Objective Functions' of 'The Lazy Bureaucrat Scheduling Problem'

This bureaucrat has a one-track mind. His objective, as Arkin, Bender, Mitchell, and Skiena describe it, is: 'to minimize the amount of work he does (he is "lazy"). He is subject to a constraint that he must be busy when there is work that he can do; we make this notion precise ... The resulting class of "perverse" scheduling problems, which we term "Lazy Bureaucrat Problems", gives rise to a rich set of new questions.'

Other mathematicians and computer scientists took their own whacks at managing lazy bureaucrats.

Arash Farzan and Mohammad Ghodsi at Sharif University of Technology in Tehran presented a paper in 2002 at the Iran Telecommunication Research Center. Titling it 'New Results for Lazy Bureaucrat Scheduling Problem', they announced that, in a mathematical sense, lazy bureaucrats are nearly impossible to manage well. A good solution, they said, is even 'hard to approximate'. What they meant: no one could say for sure that the problem could be solved – even if someone works at it ceaselessly until the end of time.

In 2003, Ghodsi and two other colleagues presented a new study. What would happen, they asked, if one imposed some tighter constraints on the lazy bureaucrat? The answer: the problem would be only slightly less nearly impossible to manage, even in theory.

These and other studies at least demonstrate that annoying people, some of them, can be described mathematically. And that on paper (or in a computer), there might be better – although not necessarily good – ways to manage them.

Managing a problem, though, does not necessarily solve it.

The mathematicians who tackle these lazy bureaucrat problems take the lazy approach. None does the hard work necessary to actually solve the problem – they give no advice about getting rid of the lazy bureaucrats. Like most non-mathematicians, they let the lazy bureaucrats career on, forever clogging the system.

For a hard worker, to read these studies is to take a descent into maddeningness.

But not everyone feels that way. The Royal Economic Society issued a press release in 2008 bearing the headline 'Lazy Bureaucrats: A Blessing in Disguise'. Touting a study by Josse Delfgaauw and Robert Dur at Erasmus University, Rotterdam, the Royal Society says: 'Hiring lazy people into the civil service helps to keep the cost of public services down'. The study itself is, as the saying goes, more nuanced.

Arkin, Esther M., Michael A. Bender, Joseph S. B. Mitchell, and Steven S. Skiena (1999). 'The Lazy Bureaucrat Scheduling Problem.' *Algorithms and Data Structures* 1663: 773–85.
Farzan, Arash, and Mohammad Ghodsi (2002). 'New Results for Lazy Bureaucrat Scheduling Problem.' *Proceedings of the 7th CSI Computer Conference,* Iran Telecommunication Research Center, Tehran, 3–5 March 2002: 66–71.
Esfahbod, Behdad, Mohammad Ghodsi, and Ali Sharifi (2003). 'Common-Deadline Lazy Bureaucrat Scheduling Problems.' *Algorithms and Data Structures: Proceedings of the 8th International Workshop, WADS,* Ottawa, Canada, 30 July–1 August: 59–66.
Gai, L., and G. Zhang (2008). 'On Lazy Bureaucrat Scheduling with Common Deadlines.' *Journal of Combinatorial Optimization* 15 (2): 191–99.

RANDOM PROMOTION DISCOVERIES

Three Italian researchers were awarded the 2010 Ig Nobel Prize in management for demonstrating mathematically that organizations would become more efficient if they promoted people at random. But their research was neither the beginning nor the end of the story of how bureaucracies try – and fail – to find a good promotion method.

Alessandro Pluchino, Andrea Rapisarda, and Cesare Garofalo of the University of Catania, Sicily, calculated how a pick-at-random promotion scheme compares with other, more enshrined methods. They gave details in a report published in the journal *Physica A: Statistical Mechanics and its Applications*.

Pluchino, Rapisarda, and Garofalo based their work on the Peter Principle – the notion that many people are promoted, sooner or later, to positions that overmatch their competence.

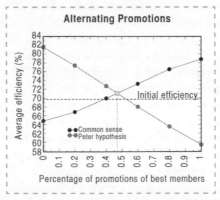

The centre point – the intersection of common sense and the Peter Principle – gives 'the most convenient strategy to adopt if one does not know which mechanism of competence transmission is operative in the organization'. Adapted from 'The Peter Principle Revised: A Computational Study'.

The three cite the works of other researchers who had taken tentative, exploratory steps in the same direction. They fail, however, to mention an unintentionally daring 2001 study by Steven E. Phelan and Zhiang Lin, at the University of Texas at Dallas, that was published in the journal *Computational & Mathematical Organization Theory*.

Phelan and Lin wanted to see whether, over the long haul, it pays best to promote people on supposed merit (we try, one way or another, to measure how good you are), or on an 'up or out' basis (either you get promoted quickly or you get the boot), or by seniority (live long and, by that measure alone, you will prosper). As a benchmark, a this-is-as-bad-as-it-could-possibly-get alternative, they also looked at what happens when you promote people at random. They got a surprise: random promotion, they admitted, 'actually performed better' than almost every alternative. Phelan and Lin seemed (at least in my reading of their paper) almost shocked, even intimidated, by what they found.

But where Pluchino, Rapisarda, and Garofalo would later, independently, hone and raise this discovery for the world to admire, Phelan and Lin merely muttered, ever so quietly in the middle of a long paragraph, that 'this needs to be further investigated in our future studies'. Then, by and large, they moved on to other things.

Human beings, many of them, are clever. Always there is potential to devise a new, perhaps better method of choosing which individuals to promote in an organization. More recently, Phedon Nicolaides, of the European Institute of Public Administration Maastricht, the Netherlands, suggested what he sees as an improvement on random promotion: randomly choose the people who will make the promotion decisions. Professor Nicolaides published his scheme in the *Cyprus Mail* newspaper.

Another, very different, non-random method was devised for use by the US Air Force. Details appear in a 170-page paper prepared in 2008 by Michael Schiefer, Albert Robbert, John Crown, Thomas Manacapilli, and Carolyn Wong of the Rand Corporation. But regardless of its merits, this Air Force scheme may be doomed to rejection purely because it has a curious name. The report is called 'The Weighted Airman Promotion System'.

Pluchino, Alessandro, Andrea Rapisarda, and Cesare Garofalo (2010). 'The Peter Principle Revisited: A Computational Study.' *Physica A: Statistical Mechanics and its Applications* 389 (3): 467–72.
Phelan, Steven E., and Zhiang Lin (2001). 'Promotion Systems and Organizational Performance: A Contingency Model.' *Computational & Mathematical Organization Theory* 7: 207–32.
Schiefer, Michael, Albert A. Robbert, John S. Crown, Thomas Manacapilli, and Carolyn Wong (2008). 'The Weighted Airman Promotion System. Standardizing Test Scores.' Rand Corporation report prepared for the US Air Force, http://www.dtic.mil/cgi-bin/GetTRDoc?AD=ADA4854 97&Location=U2&doc=GetTRDoc.pdf.

ELEVEN

FOR DETECTIVES

MAY WE RECOMMEND

Original Communication

Are full or empty beer bottles sturdier and does their fracture-threshold suffice to break the human skull?

Stephan A. Bolliger MD (Senior Forensic Pathologist)*, Steffen Ross MD (Radiologist), Lars Oesterhelweg MD (Forensic Pathologist), Michael J. Thali MD (Professor, Director, Forensic Pathologist), Beat P. Kneubuehl PhD (Physicist)

ARTICLE INFO

ABSTRACT

Article history:
Received 20 June 2008

Beer bottles are often used in physical disputes. If the bottles break, they may give rise to sharp trauma. However, if the bottles remain intact, they may cause blunt injuries. In order to investigate whether full

'ARE FULL OR EMPTY BEER BOTTLES STURDIER AND DOES THEIR FRACTURE-THRESHOLD SUFFICE TO BREAK THE HUMAN SKULL?' by S. A. Bolliger, S. Ross, L. Oesterhelweg, M. J. Thali, and B. P. Kneubuehl (published in the *Journal of Forensic and Legal Medicine*, 2009, and honoured with the 2009 Ig Nobel Prize in peace)

Some of what's in this chapter: Hair, officially and forensically • Toilet graffiti stakeout • Net-trapping bank robbers • One way to eject a hijacker • Old monk breath • Fat criminals • A new look at knifing • Psychotic security guards • O, O, O, O, O • Ethicists who steal books

FBI AGENT HAIR GUIDE

Despite its reputation for sporting nearly identical conservative hair-cuts, the FBI – the Federal Bureau of Investigation, America's government gumshoes – assembled and published an all-inclusive guide to hair. And despite its reputation for tight-lippedness, it made its guide available to anyone who might have a use or desire for it.

'Hair Bibliography for the Forensic Scientist' might make a fine gift for anyone who cares about the sometimes-tangled relationship between hair and crime. You might regard this as completely legal intellectual pornography for those who watch *CSI* (for: Crime Scene Investigation).

The author, Max Houck of the FBI's Trace Evidence Unit in Washington, DC, at least pretended that his seventeen-page report, published in 2002 in the journal *Forensic Science Communications*, would appeal strictly to professionals. 'It is hoped that this listing will provide some assistance to forensic hair examiners who are seeking information and support for courtroom forensic challenges,' he wrote.

Some light, even playful, touches suggest that Houck knew his paper would also find its way to amateurs, and even to casual fans of the hair-and-crime game. The list begins coyly with 'Don't Miss a Hair', a seven-pager published in 1976 in the *FBI Law Enforcement Bulletin*. Next comes more cuteness: 'From Bad to Worse: Hair Today, Scorned Tomorrow', which you can find in a 1997 issue of the journal *Science Sleuthing*. Then appears a more specific, and mildly grisly, title that's worded with strange ambiguity: 'Laboratory Solves Variety of Crimes with Animal Hairs' (*FBI Law Enforcement Bulletin*, 1960). Then a return to the strictly human, and a higher degree of specificity: 'Pigmentation in a Central American Tribe with Special Reference to Fair-headedness' (*American Journal of Physical Anthropology*, 1953).

One report's title almost begs you to go and find out how, exactly, its authors, D. L. Exline, F. P. Smith, and S. O. Drexler, gathered their knowledge: 'Frequency of Pubic Hair Transfer During Sexual Intercourse', published in the *Journal of Forensic Sciences*, 1998.

The list is dominated by ten items written or co-written by B. D. Gaudette. That's Barry D. Gaudette, chief scientist for hair and fibre of the Royal Canadian Mounted Police. Gaudette's 'An Attempt at Determining Probabilities in Human Scalp Hair Comparison', co-written with E. S. Keeping in the *Journal of Forensic Sciences* in 1974, specifies a rule of thumb that you can use in writing either a crime novel or a mathematics textbook: 'It is estimated that if one human scalp hair found at the scene of a crime is indistinguishable from at least one of a group of about nine dissimilar hairs from a given source, the probability that it could have originated from another source is very small, about 1 in 4,500. If, instead of one hair, n mutually dissimilar human scalp hairs are found to be indistinguishable from those of a given source, this probability is then estimated to be (1/4,500) to the nth power, which is negligible when n is greater than or equal to three.'

Houck, Max (2002). 'Hair Bibliography for the Forensic Scientist.' *Forensic Science Communications* 4 (10): http://www.fbi.gov/about-us/lab/forensic-science-communications/fsc/jan2002/houck.htm/.

Gaudette, B. D., and E. S. Keeping (1974). 'An Attempt at Determining Probabilities in Human Scalp Hair Comparison.' *Journal of Forensic Sciences* 19 (3): 599–606.

Exline, D. L., F. P. Smith, and S. O. Drexler (1998). 'Frequency of Pubic Hair Transfer During Sexual Intercourse.' *Journal of Forensic Sciences* 43 (3): 505–8.

THE WATCH ON THE LOO

In 1992 a professor named T. Steuart Watson discovered a completely effective way to prevent people from writing on public toilet walls. Then at Mississippi State University and now a professor at Miami University of Ohio, Watson published his report in the *Journal of Applied Behavior Analysis*. In it, he describes both his method and the relentless manner in which he tested it.

He carried out the experiment in three men's toilets. Each chamber had a history writ large, and small, in many different hands. The study states that 'during the preceding months, each of the walls had been repainted numerous times due to the proliferation of graffiti'.

Each day, Watson and his minions meticulously counted how many marks were on each wall. They tallied each letter, number, or piece of punctuation. Other shapes called for special assessment. The study describes one typically difficult example: 'A drawing of a happy face was counted as five marks (one for each eye, one for the nose, one for the mouth, and one for the circle depicting the head).'

The investigators employed professional stealth. 'During observations,' the report stipulates, 'only one observer entered the restroom at a time, and if another person entered to use the facilities, the observer discontinued counting and waited until the bathroom was empty before resuming counting.'

New graffiti popped up every day, in every one of the bathrooms.

But 'after treatment was implemented', Watson reveals, 'no marking occurred on any of the walls, and they remained free of graffiti at 3-month follow-up'. No marking at all. None. Not a jot. Cleanliness uninterrupted. This was complete, utter success.

The treatment was simple: 'Taping a sign on the wall that read, "A local licensed doctor has agreed to donate a set amount of money to the local chapter of the United Way [a heavily publicized American charity organization] for each day this wall remains free of any writing, drawing, or other markings".'

'The doctor', reveals the study, 'was the author, a licensed psychologist, and the amount of money donated was 5 cents per day per bathroom.'

The study lasted fifty days. Thus, with three public bathrooms in play, the maximum total potential payoff for charity was $2.50 (£2.30) per location – an aggregate $7.50 if no one ever made a mark on any wall in any of them.

Why was the treatment so very – nay, completely – effective? Watson speculates that 'prior to posting the signs, bare walls appeared to function as discriminative stimuli for graffiti, perhaps because it was not apparent that anyone cared. Posting the signs was evidence that a prominent citizen (a doctor) was prepared to pay for results'

'An alternative explanation,' he says, 'is that the presence of the observers prompted restroom users to refrain from writing on walls.'

Watson, T. Steuart (1996). 'A Prompt Plus Delayed Contingency Procedure for Reducing Bathroom Graffiti.' *Journal of Applied Behavior Analysis* 29 (1): 121–24.

CROSSWORDS AND LINEUPS

Crossword puzzles are a threat to the criminal justice system. Indeed, they may have been doing damage for decades, causing guilty persons to be set free and innocent ones to become enmeshed in hellish entanglements with the courts and jails. A 2006 study by Michael B. Lewis, a senior lecturer at Cardiff University, published in the journal *Perception*, reveals that the danger comes mostly from one variety of crossword puzzle.

Lewis has no qualms identifying the culprit. Beware, he warns, of the so-called cryptic crossword puzzle. Accordingly, the study is called 'Eye-witnesses Should Not Do Cryptic Crosswords Prior to Identity Parades'.

Once you know what to look for, cryptic crosswords are easy to recognize. The regular, or 'literal', crossword, Lewis writes, 'is a task where words must be filled within a grid where the clues to these words are literal definitions'. Cryptic crosswords 'use a similar grid but the clues involve double meanings and sometimes involve anagrams or uncommon ways of thinking about words'.

Cryptic crosswords enter the picture in seemingly innocuous ways. Police or court officials may – through a toxic mix of good intentions and ignorance – be tempted to introduce them exactly where they can do harm. Lewis explains: 'The identification of an offender by a witness to a crime often forms an important element of a prosecution's case. While considerable importance is placed by jurors on the identification of the offender by a witness (such as a suspect being picked out from an identity parade), research tells us that these identifications can often be wrong and sometimes lead to wrongful convictions.'

'It would be undesirable', he writes, 'to have witnesses doing something before an identity parade that would make them worse at picking out the offender ... Consider what witnesses may do before an identity parade. It is possible that they might be doing something to pass the time (eg read or do a puzzle). It is possible that some of these potential activities may lead to a detriment in face processing.'

Determined to determine whether reading or doing a puzzle can lead to a detriment in face processing, Lewis did an experiment. In his words: 'The tasks tested within the experiment presented here were: reading a passage from Dan Brown's *Da Vinci Code*; solving a sudoku puzzle; solving a literal crossword; solving a cryptic crossword'.

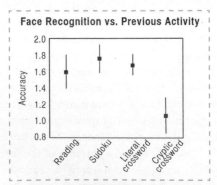

As featured in *Perception* vol. 35's 'Last But Not Least'

Sixty volunteers took part. They looked at some faces, 'then engaged in their puzzle or read the passage for 5 minutes'. Lewis then began to test their memory of the faces. 'Between each test item, however, participants continued with their puzzle or read the text for 30 seconds.'

Sudoku and literal crosswords seemed not to affect how well the volunteers identified the faces. But, according to Lewis, when the volunteers did cryptic crossword puzzles, they became less reliable at recognizing faces: 'In doing a cryptic crossword, one typically has to suppress the immediately obvious meaning of a word within the clue in favour of less obvious and more cryptic meanings. The suppression of the obvious features of the face, the obvious global letter, or the obvious literal meaning of a word may provide the device by which face-recognition performance is affected. This observation, however, does not explain how such suppression has such a detrimental effect on face recognition. That is, the question of what the mechanism is by which any of these tasks influences the supposedly modular face-recognition system is not addressed here.'

The study hammers home its message: 'The practical implication of this research is, as the title suggests, that eye-witnesses should not do cryptic crosswords before an identity parade.'

Lewis, Michael B. (2006). 'Eye-witnesses Should Not Do Cryptic Crosswords Prior to Identity Parades.' *Perception* 35: 1433–36.

DROP-OF-NET VALUE

Kuo-cheng Hsieh won the 2007 Ig Nobel Prize in economics for patenting a device that catches bank robbers by dropping a net on them. But amid the glamour of the announcement, certain of its charms may have gone unnoticed.

The invention is a nod to ancient methods of capturing animals in a forest, and also to the fanciful adaptation of those techniques in early cops-and-robbers films. Hsieh's patent sums it up in a terse ninety-four words: 'A net trapping system for capturing a robber immediately is used in a place of business such as a bank. The device looks like a storing box and is installed above the entrance of the business. When a robbery takes place and the system is activated, an infrared detecting device determines if a robber is in a zone beneath the storing box. A net, a curtain, and a plurality of barriers will drop down immediately and simultaneously. After a lifting motor is activated, the system traps the robber and suspends him above the floor.'

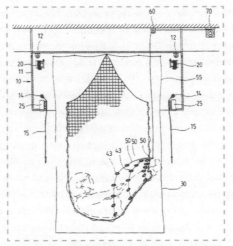

From 'Net Trapping System for Capturing a Robber Immediately'

Prior to the Ig Nobel ceremony, we were unable to locate Mr Hsieh. Attempts to contact him by telephone, by letter, and even by visits to his home address in the city of Taichung, Taiwan, all failed. Fears arose that maybe the poor man had become trapped inside his own invention.

Happily, newspaper accounts of the ceremony reached him, and he asserted his existence. We learned, via journalists in Taiwan, that Hsieh runs a security company, that he is a former commander of an amphibian frogman unit, that he once set up similar traps underwater to capture swimming Chinese spies, and that he is diligently trying to market his machinery to banks – though so far without consummating a sale.

Hsieh's patent has had at least one recorded effect: it inspired Zoltan Egeresi in the fight against terrorists.

Two years after New York City and Washington, DC, were attacked by hijacked planes, Egeresi, a California inventor, filed a patent application for an 'anti hijacking system'. In an ingenious commingling of ideas, Egeresi adapted the Hsieh bank-robber-trapping technology as a way to simplify a rather costly anti-hijacker system devised in the early 1970s by Gustano A. Pizzo.

Pizzo's invention involves some clever, prescient engineering. An area – a little waiting room, really – is partitioned off between the crew cabin and the plane's passenger section. No mere anteroom this. It's an anti-hijacking room, with a specially built mechanical floor. A hijacker, finding himself isolated in this lonely space, is at the mercy of its machinery. He will fall prey to – indeed will fall through – specially hinged floor panels. As Pizzo explains it, these pivoting panels 'are adapted to be dropped for lowering the hijacker into a releasable capsule to which a parachute is attached. Bomb bay doors are provided in the belly of the plane for ejection of the capsule from the aircraft'. The pilots control the action. Safely tucked into the command cabin, they activate an electromechanical latch. The floor panels, unlatched, suddenly give way. The hijacker falls into a small but yawning pit lined with a sturdy net. A draw-cord automatically seals the hijacker inside what is now an ovoid imprisonment capsule. The pilots can, at their leisure, choose an appropriate moment to expunge the encapsulated crook. As the patent describes this moment: 'The bomb bay doors are opened by air cylinders, permitting the capsule and its parachute to drop therethru.' At that point, the hijacker or hijackers, neatly packaged, fall into the waiting arms of authorities on the ground.

The mechanics of Egeresi's Pizzo-ian, Hsieh-ian hybrid invention are simpler and cheaper: 'When one or more person [sic] is trying to over power the pilots, this anti hijacking system can provide a non-lethal last line of defense. Doors on the cockpit may not be penetration proof. When pilot or flight attendant is confronted with a situation where the pilot's door is about to be penetrated, a concealed stainless steel net from below the carpet will hoist up all people to the ceiling.'

Egeresi is not alone in taking inspiration from Pizzo's patent. In the wake of the 9/11 attacks on New York City, sixteen inventors filed patent applications making reference to Pizzo's earlier work.

Hsieh, Kuo-cheng (2007). 'Net Trapping System for Capturing a Robber Immediately.' US Patent no. 6,219,959, 24 April.

Egeresi, Zoltan (2006). 'Anti Hijacking System.' US Patent no. 7,014,147, 12 March.

Pizzo, Gustano A. (1972). 'Anti Hijacking System for Aircraft.' US Patent no. 3,811,643, 21 May.

THE COST OF BEER

A study called 'Violence-Related Injury and the Price of Beer in England and Wales' offers support for actions by Her Majesty's Chancellor of the Exchequer to raise the tax on a pint of beer by one penny.

The study's authors, based at Cardiff University, in Wales, are scholars of pence and beer. Kent Matthews is the Sir Julian Hodge professor of banking and finance. Jonathan Shepherd, a professor at the dental school and director of that school's violence research group, has long campaigned for the mandatory use of non-glass 'glasses' and bottles in late-night drinking establishments. Shepherd's colleague Vaseekaran Sivarajasingham also took part.

The question 'What causes violent behaviour?' is not simple. The trio relate some insights they gleaned from others' research. Apparently, people who are frequently drunk are less prone to commit violence than drinkers who are not used to being intoxicated. The study describes this in succinct technical terms: '[Those] with the lowest usual involvement with alcohol were subject to a higher elevation in their risk immediately after alcohol consumption compared to those who drank more heavily.'

Matthews, Shepherd, and Sivarajasingham caution that it can be difficult to establish the specific causes of violence. A 1994 study in the *Annals of Emergency Medicine*, which they do not mention, demonstrated this very point. Entitled 'Impact of Yankee Stadium Bat Day on Blunt Trauma in Northern New York City', it said: 'The distribution of 25,000 wooden baseball bats to attendees at Yankee Stadium did not increase the incidence of bat-related trauma in the Bronx and northern Manhattan. There was a positive correlation between daily temperature and the incidence of bat injury. The informal but common impressions of emergency clinicians about the cause-and-effect relationship between Bat Day and bat trauma were unfounded.'

The new beer/violence analysis was fairly straightforward. Matthews, Shepherd, and Sivarajasingham looked up sets of numbers for each region in England and Wales – assault rates reported in hospital emergency departments, the local price of beer, the local unemployment rate, and other likely suspects. Then they compared region against region.

Their report concludes that 'the regional distribution of the incidence of violent injury is related to the regional distribution of the price of beer'. It predicts, quite specifically, that a one percent increase in the cost of beer would result in five thousand fewer cases of assault every year. With the UK-wide average price of a pint standing at about

£2.80 ($4.40), a penny-a-pint action might thus be expected to prevent about 1800 assaults in the coming year.

But the study supplies the logic for an intriguing, alternative method by which the government's treasurer might reduce the number of assaults: prevent young people from having jobs. As the researchers explain it, 'there is a strong negative relationship between youth unemployment and violence related injury. The higher is unemployment, the lower is youth disposable income, and the lower the consumption of alcohol and consequently the lower the incidence of violent injury.'

Matthews, Kent, Jonathan Shepherd, and Vaseekaran Sivarajasingham (2006). 'Violence-Related Injury and the Price of Beer in England and Wales.' *Injury* 37 (5): 388-94
Bernstein, S. L., W. P. Rennie, and K. Alagappan (1994). 'Impact of Yankee Stadium Bat Day on Blunt Trauma in Northern New York City.' *Annals of Emergency Medicine* 23 (3): 555–59.

SOMETHING FOUL IN THE AIR

Automated breath analysis, the policeman's best tool for identifying drunk drivers, has a new use. Three Greek chemists tell all in a report called 'Analysis of Expired Air of Fasting Male Monks at Mount Athos'. Published in a magazine few police officers ever read – the *Journal of Chromatography B* – the study describes a new reason to appreciate monks. From a scientist's point of view, fasting monks are a reasonable substitute for 'entrapped people under the ruins of a collapsed building after an earthquake'.

The report explains: 'Survivors are often trapped in voids of ruins, usually dehydrated and starved ... Expired air volatile organic compounds, along with volatiles of other biological fluids (blood, urine and sweat), might give indications of human life or loss. To study expired air under similar situations one needs to find volunteers for providing breath samples. However, identification of a group of volunteers undergoing starvation for 72 hours (crucial time for search and rescue operations) for experimentation purposes may be difficult to find.'

Difficult, yes. But not impossible. Thus: monks. The monks at the Great and Holy Monastery of Vatopaidi, on the Athos peninsula in the Aegean Sea, are famous for fasting. For three days prior to Easter, they ingest neither food nor water.

Seven monks made their breath available to the scientists. These monks preceded their fast with a Sunday evening meal of fish, salad, and wine. They ended it with a special hot soup, called *housafi*, consisting of plums, figs, grapes, oranges, and other fruits. But before slurping the soup, they did some heavy breathing into plastic collection bags.

Starved-monk breath is a treasure, so it was handled with care. The scientists pumped it from the collection bags into special tubes. They then fed the tubes into a gas chromatograph, an instrument that separated the breath into its constituent parts.

Here, in case you need to know, are the twenty-nine most frequent volatile substances in the breath of the Mount Athos monks after three days of starvation (but before they had their first taste of fruit soup): Acetone; phenol; di-limonene; 2-pentanone; isoprene; acetaldehyde; n-octyl acetate; dichloromethane; octane; hexane, 3-methyl; hexane, 2-methyl; heptane; 2-beta-pinene; heptane, 2-methyl; heptane, 4-methyl; heptane, 3-methyl; carbonic acid, dimethyl ester; heptane, 2,4-dimethyl; ethanone, 1-phenyl; benzene, 1,2,3-trimethyl; cyclohexane, methyl; cyclohexanone; benzene, (1-methylethenyl); toluene; nonane; 1-hexanol, 2-ethyl; 2-butanone; cyclohexane, 1,4-dimethyl; and benzene, 1,2-dimethyl.

The acetone dominated everything else. The report notes, rather dryly, that 'the odour of acetone was detectable by smell in the expired air of the monks'. The odour is familiar to many people who have never fasted: acetone is nail polish remover. And it's one of the substances a human body produces when it burns fat reserves rather than food.

Analysis of expired air of fasting male monks at Mount Athos

M. Statheropoulos, A. Agapiou *, A. Georgiadou

National Technical University of Athens (NTUA), School of Chemical Engineering, Sector I, 9 Iroon Polytechniou Street, Athens 157 73, Greece

Some scientists appreciate the fasting monks of Mount Athos not just for their pungent breath. That's a whole other story. You can read all about it in a 1994 report called 'An Epidemiological Study of Headache Among the Monks of Athos (Greece)'.

Statheropoulos, M., A. Agapiou, and A. Georgiadou (2006). 'Analysis of Expired Air of Fasting Male Monks at Mount Athos.' *Journal of Chromatography B* 832: 274–79.

Mitsikostas, D. D., A. Thomas, S. Gatzonis, A. Ilias, and C. Papageorgiou. (1994). 'An Epidemiological Study of Headache Among the Monks of Athos (Greece).' *Headache* 34 (9): 539–41.

T IS FOR TEMPTATION

'The purpose of this study was to examine meanings assigned by observers to an adolescent wearing an alcohol promotional T-shirt.' So begins a study published in the September 2004 issue of the *Family and Consumer Sciences Research Journal*.

Scholars had never tackled this exact question. Now they have.

These particular scholars, Jane E. Workman, Naomi E. Arseneau, and Chandra J. Ewell, are based at Southern Illinois University. The details of their discovery are numerous. Unlike many in their trade, Workman, Arseneau, and Ewell are good at baking a morass of data down into a concise, clear description. Here is their take: 'Regardless of gender, the adolescent wearing an alcohol T-shirt [rather than a plain T-shirt] was rated as less honest, less independent, less responsible, less feminine, less reliable, less religious, less likely to be on time, less likely to do well in school, more likely to smoke, more likely to be a party animal, more likely to drink, more likely to be a risk taker, and more likely to use profanity.'

The researchers do not say that alcohol T-shirt wearers are less honest or less abstemious or less anything else than other teens. The study, they insist, merely 'provides an important piece of evidence regarding perceptions associated with alcohol promotional T-shirts'.

Ah, but they do see danger. Wearing the wrong kind of shirt 'might lead peers to form impressions of the wearer as independent, autonomous, sociable, irresponsible, and a risk taker'. And that 'might lead to expanded opportunities and pressures for adolescents to engage in risky behaviors'. In other words, when you wear the shirt, you tempt others to tempt you onto the road to ruin.

Workman, Arseneau, and Ewell of course did not claim that their entire theory is pristinely original. They stand, metaphorically, on the shoulders of T-shirt giants. Those giants are Donna K. Darden and Steven K. Worden, whose 1991 report 'Identity Announcement in Mass Society: The T-Shirt', proclaimed a theory explaining how T-shirts can function as symbols. It was published to mild acclaim in the journal *Sociological Spectrum*.

Worden and Darden, by the way, are known for more than just their T-shirt work. They are, in some ways, the first couple (academically speaking) of cock-fighting. Their report 'Knives and Gaffs: Definitions in the Deviant World of Cockfighting' was published quietly in 1992, in the journal *Deviant Behavior*. Eight years later it was republished as a chapter in the surprisingly dull book *Deviance and Deviants: An Anthology*. According to the book's publisher: 'Worden and Darden argue that even within a deviant setting, some participants are defined as more deviant than others'.

Workman, Arseneau, and Ewell, in their T-shirt study, hope to prevent deviance. Their report ends with a stark, clarion call: 'School administrators sometimes need empirical evidence to justify a ban on certain clothing items. [Our] study provides empirical evidence to justify a ban on alcohol promotional clothing items.'

Workman, Jane E., Naomi E. Arseneau, and Chandra J. Ewell (2004). 'Traits and Behaviors Assigned to an Adolescent Wearing an Alcohol Promotional T-Shirt.' *Family and Consumer Sciences Research Journal* 33 (1): 498–516.

Darden, Donna, and Steven Worden (1991). 'Identity Announcement in Mass Society: The T-shirt.' *Sociological Spectrum* 11 (1): 67–79.

Worden, Steven, and Donna Darden (1992). 'Knives and Gaffs: Definitions in the Deviant World of Cockfighting.' *Deviant Behavior* 13: 271–89.

FAT CHANCE OF CRIMINALITY?

Fat people are more likely to become criminals, and their very fatness may help shape their criminality. That's the conclusion reached by Professor Gregory N. Price in a study called 'Obesity and Crime: Is There a Relationship?' published in the journal *Economics Letters*.

Price, an economist at Morehouse College in Atlanta, Georgia, writes that his findings accord with a wide body of earlier economics research: 'There is evidence that for individuals, being obese lowers wages, reduces labor force participation, constrains occupational attainment, and inhibits the formation of human capital that is important for labor market success. To the extent these labor market effects of obesity reduce the incentives an individual has for engaging in legitimate labor market activities, it is plausible obesity could increase individual incentives for engaging in illegitimate activities such as crime – an idea which we explore empirically.'

Using that traditional royal 'we', Price explains: 'Our data consists of offenders with last name starting with the letter "A" incarcerated in the state of Mississippi as of August 20, 2005.'

He (that is, 'we') took nineteen variables into consideration. These include each person's age, gender, height, waist circumference, race, and thirteen different aspects of the person's fatness.

One variable is called 'scrabble'. Price explains that 'the scrabble score of inmate first name is based upon the numeric values assigned to letters in the board game Scrabble, produced and distributed by both Mattel Inc. and Hasbro Inc'. He cites earlier studies (by researchers named Figlio, Bertrand, and Mullainathan) as to why the Scrabble value of a person's name is significant: 'Figlio (2005) shows that individuals with low socioeconomic status have a tendency to have first names with a high scrabble value ... Figlio finds that for black students in a large Florida public school district, having a surname with a high scrabble value, as part of an index of socioeconomic status, has a negative effect on test scores. As test scores are a component of human capital, this suggests that the blackness of a name as measured by the scrabble score, can have adverse labor market effects (Bertrand and Mullainathan, 2004), which could increase the probability that crime is acceptable to an individual.'

In a touch that seems literary as well as econometric, the paper also cites an economist named Gloom, who expounds on a fine point concerning the ratio of median to mean income.

Price ends by discussing the implications of his discovery. He writes: 'Public health policies successful at reducing obesity among individuals in the population will not only make society healthier, but also safer. If obesity among individuals in the population increases the probability that they will engage in criminal activities, reductions in obesity among individuals would also reduce individual crime hazards, and society's overall crime rate.'

Price, Gregory N. (2009). 'Obesity and Crime: Is There a Relationship?' *Economics Letters* 103: 149–52.
Gloom, Gerhard (2004). 'Inequality, Majority Voting, and the Redistributive Effects of Public Education Funding.' *Pacific Economic Review* 9: 93–101.

IN BRIEF

'CAUSE OF MICROBIAL DEATH DURING FREEZING IN A SOFT-SERVE ICE CREAM FREEZER'

by J. Foley and J. J. Sheuring (published in the *Journal of Dairy Science*, 1966)

BEHIND-THE-CUTTING-EDGE RESEARCH

In this era of stun guns, two-thousand-pound bombs, and too-ad-vanced-to-be-tested strategic missile defence systems, few individuals understand the effect of knife-handle shape on stabbing performance. Ian Horsfall and his colleagues are among the proud, happy band of brothers and sisters. Their report, 'The Effect of Knife Handle Shape on Stabbing Performance', makes it easy for all of us to share in this knowledge.

The team is based at the Royal Military College of Science, Cranfield University, in Swindon, UK. The report's title is a measure of the scientists' modesty: they studied not just the handle's shape, but also its size.

'The bottom line', Horsfall says, 'is that stabbing performance is almost wholly dependent on the person and not a function of the knife handle.' He emphasizes that 'this paper doesn't in any way illustrate how to stab people'. The thrust of the research is how to protect people against knife stabs, and, especially, how to design police body armour.

Previous investigations, including some by these same scientists, looked at the basic physics of stabbing. The topic was of interest to Arthur Conan Doyle, who reported the following exchange between Sherlock Holmes and Dr Watson: 'He chuckled as he poured out the coffee. "If you could have looked into Allardyce's back shop, you would have seen a dead pig swung from a hook in the ceiling, and a gentleman in his shirt sleeves furiously stabbing at it with this weapon. I was that energetic person, and I have satisfied myself that by no exertion of my strength can I transfix the pig with a single blow."'

No one, not even Holmes, gave full attention to the geometries of knife handles. Horsfall et al. sought to correct this oversight.

Knife-stabbing-force analysis is partly theoretical, partly experimental. The experiments tend to be colourful, sometimes involving the carcasses of hefty vertebrates and some gaudily sharp instruments (sabers and 'commando' style blades were once in vogue). As Horsfall et al. put it in an earlier paper, stabbing 'is a relatively complex task, as not only are there a wide variety of possible weapons but also a highly variable human element in powering the weapon'.

The new experiment used eleven volunteer stabbers of various sizes. They used a knife that was fitted with special measurement instruments, and at various times, with four interchangeable – and very different – kinds of handle.

'Underarm stab action' demonstrated

The volunteers did not get to stab any carcasses. Instead, they plunged their blades into a meat substitute – a 5.5-millimetre-thick aramidthermoplastic composite sheet held in front of a large clay block. After all was done and said, they had demonstrated that fussing about knife handles is a nearly irrelevant, albeit gripping, distraction. No matter how you slice it, he stabs best who is skilful and strong.

Horsfall, I., C. Watson, S. Champion, P. Prosser, and T. Ringrose (2005). 'The Effect of Knife Handle Shape on Stabbing Performance.' *Applied Ergonomics* 36 (4): 505–11.
Horsfall, I., P. D. Prosser, C. H. Watson, and S. M. Champion (1999). 'An Assessment of Human Performance in Stabbing.' *Forensic Science International* 102 (2-3): 79–89.

CRAZED SECURITY

'Who watches the watchers?' becomes an especially interesting question when the watchers are psychotic. A team of doctors from Texas and California explored this question in 1993 in the *Journal of Forensic Sciences*. Their names are J. A. Silva, G. B. Leong, and R. Weinstock. Their study is called 'The Psychotic Patient as Security Guard'. Quickly, they confide: 'Although the public assumes that only mentally healthy individuals who possess the capability to handle stressful situations become employed as security guards, this may not be the case.'

The doctors discuss what they call 'a small sample of security guards who suffered from psychotic disorders'. That phrase 'a small sample' is tantalizing. It suggests the unasked and unanswered question: 'What percentage of security guards are psychotic?' The authors do, somewhat disjointedly, make a point of saying that 'further work is needed to determine the proportion of psychotic inpatients who are security guards'.

Their small sample consists of fifteen guards, one of whom, identified as Mr A, gets an especially detailed examination. Mr A became a security guard after being discharged from prison for using a knife to try to obey voices that told him to kill a stranger. The doctors report that eventually 'Mr A quit his security guard job on the advice of his auditory hallucination'.

Thirteen of the fifteen met all the diagnostic criteria for paranoid-type schizophrenia. The other two were classed as having schizo-affective disorder. Eight said they had experienced hallucinations and paranoid delusions while on the job.

Of the fifteen, only eight mentioned having a history of aggression, and only three said they had attacked people with a knife. Only one had assaulted someone while on the job; he was the only one of the fifteen who routinely carried a weapon while at work. Two others said they had obtained permits to carry guns, but had not taken advantage of those permits.

This particular combination of profession and mental condition, the report hints, may have its tiny good side. 'A mild degree of suspiciousness', they write, 'may be adaptive for the security guard.' This is illustrated by the case of Mr A. Although Mr A became increasingly paranoid during his time on the job, 'he felt safe as his auditory hallucinations would warn him and help him discriminate potential intruders from passers-by'.

The doctors say that all fifteen psychotic guards cited monetary motivation as their primary reason for taking up their line of work. The doctors do not indicate whether or how this differs from the motivation of non-psychotic security guards.

The report advises mental health professionals 'to collect a work history, including security guard work, among psychotic patients'. It advises the public, in mildly tangled language, that 'psychotic security guards carrying weapons may be the greatest risk of posing a danger to others'. And for employers, there is this: 'The question of who should be employed as security guards, especially if weapons will be carried, merits further study.'

Silva, J. A., G. B. Leong, and R. Weinstock (1993). 'The Psychotic Patient as Security Guard.' *Journal of Forensic Sciences* 38 (6): 1436–40.

0?

'What's in a name?' Shakespeare famously asked – but from a detective's viewpoint, the question is dauntingly broad. A team of scientists based in Switzerland and France undertook a more tightly targeted investigation: what's in a capital letter O? They spill all in a study published in the journal *Forensic Science International*.

This was their way of tackling a deep legal worry. Police and other criminal justice authorities struggle, on occasion, to decide the significance of a sample of handwriting. These professionals rely on the two so-called fundamental laws of handwriting: first, that no two people write exactly alike; and second, that no one person writes the same word exactly the same way twice. The problem is that no one knows whether these 'laws' are correct. Maybe, just maybe, our system of jurisprudence rests on assumptions that are, like the letter O itself, hollow.

Raymond Marquis at the University of Lausanne's School of Criminal Sciences, together with three colleagues, took an unflinching look at this possibly gaping hole in the legal system. They examined handwriting samples from three individuals. Collectively, the samples contained 445 handwritten capital Os that the scientists deemed suitable for analysis.

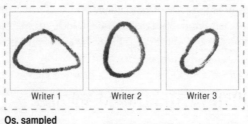

Writer 1   Writer 2   Writer 3

Os, sampled

TV crime dramas have given us a misleading notion about handwriting analysis. The state of the art is substantially that – a traditional art, larded with some nice dollops of rigorous science. Marquis and his team write: 'Letter shape has not been studied in a global and precise way within the various existing methods; only certain aspects of it have been approached by a variety of geometrical measurements.'

It's true. No other handwritten uppercase letter has been studied with exactly the same computational rigour that Marquis's team applied to their chosen letter. Not Y, not M, not C. Not even A.

The team stripped down each capital O, step by step. First, they digitized the O, turning it into a mass of data that would be digestible by any healthy computer. Then they removed the fat from the written lines, paring each particular O down to its own, peculiar, skinny, skeletal shape – a wiggly, quirky contour. They then performed a Fourier analysis, which is a sort of mathematical X-ray. This revealed a series of simple pictures, each showing some simple aspect of that particular O's personal O-ness – its ellipticality, its triangularity, its quadrangularity, its pentagonality, its hexagonality. Together, these give a good, round, precise picture of that uniquely individual O.

The scientists are excited at what they found. The Fourier analysis did indeed reliably tell them which of the 445 capital Os had been handwritten by which of the three people.

They have not yet answered the big question: can we trust the laws of handwriting analysis? But they've got things moving in the right direction. Marquis and his colleagues have achieved, they say pointedly, 'a step to objective discrimination between writers based on the study of the capital character O'.

Marquis, Raymond, Matthieu Schmittbuhl, Williams David Mazzella, and Franco Taroni (2005). 'Quantification of the Shape of Handwritten Characters: A Step to Objective Discrimination Between Writers Based on the Study of the Capital Character O.' *Forensic Science International* 150 (1): 23–32.

DO ETHICISTS STEAL MORE BOOKS?

Do ethicists steal more books?

Eric Schwitzgebel

If explicit cognition about morality promotes moral behavior then one might expect ethics professors to behave particularly well. However, professional ethicists' behavior has never been empirically studied. The present research examined the rates at which ethics

'One might suppose that ethicists would behave with particular moral scruple', begins the little monograph, looking you straight in the eye while grinning and snorting, textily. The two co-authors, philosophy professors who specialize in ethics, thus embark on what they call a 'preliminary investigation' of their fellow ethics experts.

Eric Schwitzgebel, of the University of California, Riverside, and Joshua Rust, of Stetson University in Deland, Florida, surveyed almost three hundred attendees of a meeting of the American Philosophical Association. Tell us, they asked in a variety of ways, about the ethical behaviour of ethicists you have known. Schwitzgebel and Rust offered candy to anyone who agreed to complete the survey form. They report that 'a number of people stole candy without completing a questionnaire or took more than their share without permission'.

The ethics experts in aggregate indicated that in their experience, on the whole, ethicists behave no more ethically than do other persons. The paper, published in the journal *Mind*, pauses for just a moment to suggest a broader context. 'Police officers commit crimes', it says. 'Doctors smoke. Economists invest badly. Clergy flout the rules of their religion.'

Schwitzgebel also wrote a study, on his own, called 'Do Ethicists Steal More Books?', which elbowed its way into the face of readers of the journal *Philosophical Psychology*. He drew up lists of philosophy books – some specifically about ethics, others not. Then, using information available through computer networks, he examined the status of every copy of those books in nineteen British and thirteen American academic library systems.

Schwitzgebel looked separately at what happened to newish books (Buchanon's *Ethics, Efficiency and the Market*; Baron's *Kantian Ethics Almost Without Apology*; Hurd's *Moral Combat*; and suchlike bestsellers), and to older ones (Aristotle's *Nicomachean Ethics*; Kant's *Critique of Judgment*; Nietzsche's *Beyond Good and Evil*; and other beloved masterworks).

It was roughly the same story. The ethics books, whether youthful or aged, went missing more often than did the not-quite-so-relentlessly-about-ethics books.

The youthful, 'relatively obscure, contemporary ethics books of the sort likely to be borrowed mainly by professors and advanced students of philosophy were actually about 50% more likely to be missing'. The aged, 'classic (pre-1900) ethics books were about twice as likely to be missing'. (For those older books, Schwitzgebel looked only at the American libraries, muttering that 'the British library catalog system proved impractically unwieldy'.)

More recently, Schwitzgebel has written in his blog about what he calls 'the phenomenology of being a jerk'. He identifies two important components of jerkhood. 'First: an implicit or explicit sense that you are an "important" person.' 'Second: an implicit or explicit sense that you are surrounded by idiots.'

To determine whether you yourself might be a jerk, Schwitzgebel suggests, look at those two simple criteria. He adds the almost mandatory thought: 'I can't say that I myself show up as well by this self-diagnostic as I would have hoped.'

Schwitzgebel, Eric, and Joshua Rust (2009). 'The Moral Behavior of Ethicists: Peer Opinion.' *Mind* 118: 1043–59.
— (2010). 'Do Ethicists and Political Philosophers Vote More Often Than Other Professors?' *Review of Philosophy and Psychology* 1: 189–99.
Schwitzgebel, Eric (2009). 'Do Ethicists Steal More Books?' *Philosophical Psychology* 22 (6): 711–25.

TWELVE

IT MUST MEAN ...
SOMETHING

IN BRIEF

'DESCARTES AND THE GUT: "I'M PINK THEREFORE I AM"'
by D. G. Thompson (published in *Oui*, 2001)

Some of what's in this chapter: Foul words for referees. • The decline of public insult in London • How beats the poet's heart, electrically • The the the the the the the the the the the, in order • The scholar's colon • Bad highlighting • Dude • Bob, by the look of him • Gówsü; Déznep; Wítaw; Thôbonf; Mávquawpûnt; Stisk • 'Meaning' meaning 'meaning'

THE CURSE OF THE REFEREE

Do swear words have predictable effects on football referees? A team of Austrian scientists tackles that question in a study called 'May I Curse a Referee? Swear Words and Consequences'. Stefan Stieger, of the University of Vienna, together with Andrea Praschinger and Christine Pomikal, who describe themselves as 'independent scientists', published their report in the *Journal of Sports Science and Medicine*.

Football referees enforce the laws of the game set forth by the sport's governing organization, FIFA (the Fédération International de Football Association). The pertinent regulation is FIFA's Law 12 ('Fouls and Misconduct'), whose very last section – Section 81 – simply says: 'A player who is guilty of using offensive, insulting or abusive language or gestures must be sent off.'

Stieger, Praschinger, and Pomikal performed their research in two steps. First, they obtained some swear words. Then, obscenities in hand, they found some referees who were willing to answer a survey.

The team began by drawing up a list of one hundred potential swear words. They pared the list by recruiting thirteen German-speaking residents of Austria, six women and seven men. Each Deutsch-Lautsprecher evaluated each word, rating both its degree of insultingness and whether it could be properly applied to both men and women. 'Participants [also] had to rate the insulting content of each swear word. Does the swear word concern the person's power of judgment (e.g., blind person), intelligence (e.g., fool), appearance (e.g., fatso), sexual orientation (e.g., bugger), or genitals (e.g., crap)?'

The researchers then found 113 game game referees from across Austria, and posed the following situation to each of them: during a stoppage in play, one team's captain comes up to you and suggests you make a particular ruling. You decline. 'Hereupon the team captain says … (the swear word mentioned below), turns around and walks [away].' Do you, the referee, respond by issuing (1) a red card or (2) a yellow card or (3) an admonition, or do you (4) do nothing at all? The referee was asked this for each of the twenty-eight swear words.

Their answers showed a clear pattern. 'Analyzing all swear words independent of their offensive nature, it was found that 55.7% of the swear words would have received a red card, although Law 12 would have prescribed a red card in all cases.' Only a very few officials would always, automatically, eject the player.

Digging into the nitty-gritty of the data, the researchers gained two general insights. First, 'that the decision to assign any card was dependent on the insulting content of the swear word'. Second, that 'referees would have issued a red card for sexually inclined words or phrases rather than for terms insulting one's appearance'.

Praschinger, Andrea, Christine Pomikal, and Stefan Stieger (2011). 'May I Curse a Referee? Swear Words and Consequences.' *Journal of Sports Science and Medicine* 10: 341–45.

MAY WE RECOMMEND

'SWEARING AS A RESPONSE TO PAIN'

by Richard Stephens, John Atkins, and Andrew Kingston (published in *Neuroreport*, 2009, and honoured with the 2010 Ig Nobel Prize in peace)

INJURY TO INSULT

Insults just aren't what they used to be, according to a study called 'The Decline of Public Insult in London 1660–1800'. The study's author, Robert B. Shoemaker, teaches eighteenth-century British history at Sheffield University, in the UK.

Professor Shoemaker pored through records of court proceedings from the late sixteenth through the early nineteenth centuries, paying special attention to the insults. Time was, insulting someone in public – or even in private – could easily propel you into court, and thence, if the insult was good or your luck wasn't, to jail.

	Specific Words	General Words	Common Disturber	Total
1660s	17	0	13	30
1670s	13	21	37	71
1680s	15	8	22	45
1690s	5	6	15	26
1700s	4	2	10	16
1710s	1	4	17	22
1720s	5	26	2	33
1730s	3	21	1	25
1740s	7	19	2	28
1750s	1	2	2	5
1760s	0	2	1	3
1770s	0	0	0	0
Total	71	111	122	304

Recognizances for Insulting Words: Middlesex Sessions Sample 1660–1780

Shoemaker charted the number of insult-fuelled prosecutions in the consistory court of London over those centuries. 'The pattern is clear', he writes, 'a massive increase in the late sixteenth century to a peak in the 1620s and 1630s, followed by a collapse ... By the late eighteenth century per capita prosecutions in London had fallen to only one or two per 100,000 per year.' By the late 1820s, the number of prosecutions had dropped to an insulting one or two, total, per year.

(The high point for legal action, by the way, was 1633, the year Samuel Pepys was born. One can only speculate at how much more colourful his famous diary might have been had Pepys lived a generation earlier, during London's golden age of insult.)

As the years rolled by, individual nasty words lost some of their power to trigger prosecution. Legal proceedings dealt, instead, with more general allegations. Court documents became less fun to read, with fewer bold, juicy epithets, the accusations now built of mushy phrases such as 'opprobrious names', 'scandalous abuse', or 'grossly insulting'.

The seventeenth and eighteenth centuries revolutionized the legal handling of insult, Shoemaker suggests in telling us that 'the very nature, function and significance of the insult was changing over this period'.

He invokes the words of King's College London historian Laura Gowing. Gowing emphasized that in earlier years, 'Defamations rarely happened inside private houses, at meals, or within private conversation, but were staged, often in the open, with an audience provided by the witnesses who, "hearing a great noise" in the street, left their work or houses to investigate or intervene ... the doorstep was a crucial vantage point for the exchange of insult.'

But by the eighteenth century, Shoemaker reports, 'the insult became less public'. Insults moved indoors. Many 'took place in semi-private locations, such as yards, shops, pubs and houses, where there were not always many witnesses'. Also, 'there was much less certainty about whether defamatory words automatically destroyed reputations', and so, 'correspondingly, the power of insulting words was declining'.

All this tells us something sad about modernization: 'At a basic level, due to frequent geographical mobility eighteenth-century

Londoners did not know or take an interest in the activities of their neighbours as much as they used to.'

In this view of things, public insults declined because modern citizens no longer loved their neighbours.

Shoemaker, Robert B. (2000). 'The Decline of Public Insult in London 1660–1800.' *Past and Present* 169 (1): 97–131.

MEASURES OF POETRY

Poetry is said, by poets, to make the heart flutter and the breath catch. A team of German, Swiss, and Austrian scientists showed that the claim is quite true, at least under certain laboratory conditions.

The researchers tried to describe this lyrically. They sought, they say, 'to investigate the synchronization between low frequency breathing patterns and respiratory sinus arrhythmia (RSA) of heart rate during guided recitation of poetry'.

Twenty healthy individuals volunteered to spend twenty minutes reading aloud hexameter verse from ancient Greek literature. These volunteers were German speakers. They read a passage from a German translation of Homer's heart-pounding, breath-forcing epic *The Odyssey*. In accordance with modern sensibilities, the research protocol was approved beforehand by an ethics committee. The study was published in the *American Journal of Physiology – Heart and Circulatory Physiology*.

Dr Dirk Cysarz, of the Gemeinschaftskrankenhaus Herdecke in Germany, led the team. He loves studying matters of the heart and lungs, especially the ways in which those organs exhibit rhythm and pacing. Other group members have spent their careers delving, variously, into the mysteries of mathematics, music, and speech problems.

To any volunteer who was unversed in the methods of modern medical experimentation, the recitation would have seemed unexpectedly complex. In ancient Greece, reciting poetry was a simple process. One stood or sat, unencumbered, and spoke. But here, now, there were strings attached. The Greeks, anyway, would have called them strings. We call them electrical wires.

Whilst spouting Homer from the lips, each volunteer was also sending electrical signals straight from his or her heart, via a transducer and wires, to a solid-state electrocardiogram recording apparatus.

And that's not all. The poetry-reciting, electrical-pulse-generating volunteer also supplied streams of information about his or her nasal and oral airflow. Three thermistors were mounted next to the nostrils and in front of the mouth. Thermistors are little electronic devices that measure temperature change – in this case between warm, exhaled air and cooler, about-to-be-inhaled air. Thus were the nuances of breath and pulse documented, forming a record of each volunteer's poetico-physiological experience.

The scientists gathered up a potentially blooming, buzzing confusion of data. To make sense of it, they used statistical and other mathematical tools that, for the most part, did not exist in the time of Homer: time-series band-pass filtering; Fourier transforms; Hilbert transforms; RR-tachograms.

The result of all this is summed up in the title of their study: 'Oscillations of Heart Rate and Respiration Synchronize During Poetry Recitation'. Though expressed in somewhat technical language, it accords with the belief of millennia of declaiming poets. The synchronization, we now know, is not perfect. But the project brings us ever-so-slightly closer to understanding poetry, inspiration, and exhalation, by the numbers.

Cysarz, Dirk, Dietrich von Bonin, Helmut Lackner, Peter Heusser, Maximilian Moser, and Henrik Bettermann (2004). 'Oscillations of Heart Rate and Respiration Synchronize During Poetry Recitation.' *American Journal of Physiology – Heart and Circulatory Physiology* 287: H579–87.

WHERE THE —

'The' has its place. That, more or less, is the theme of Glenda Browne's treatise entitled 'The Definite Article: Acknowledging "The" in Index Entries'.

The 'The' article appears in *The Indexer*, the information- and fun-packed publication for professional indexers everywhere. *The Indexer* has its own index, which includes an entry for Browne, Glenda.

Browne characterizes herself as an Australian freelance indexer. Her study is a four-page guide for the definitely perplexed. It explains: 'If "The" exists in a name or title, it should exist in the index entry for that name or title. And if it exists in the index entry, it should be taken into account when sorting the entries.'

The problem is widespread, and although there are rules (at

least three different – and differing – official sets of rules), indexers often go their own ways. Browne gives examples. In the 2000/2001 Sydney telephone directory, '"Agency Register The" and "Agency Personnel The" are sorted under "A", while "The Agency Australia" is sorted under "The". "The Sausage Specialist" is filed under "The" and "Sausage", while "The Meat Emporium" is only filed under "Meat Emporium The".'

Browne says, '"The" often doesn't matter. There are many titles that include "The", but then treat it as if it doesn't exist. The masthead of [the newspaper] *The Australian*, for example, has a tiny "The" above a large "Australian". Their layout tells us that The is insignificant, but they won't follow this through and omit it entirely. Corporate names such as "The University of Queensland" are used at times with, and at times without, an initial "The". This makes it very difficult for users to know whether "The" is an integral part of the name.'

'On the other hand,' she continues, 'in many corporate names "The" has been deliberately chosen as the first word of the name, and is used consistently. The musical group "The Beatles" is referred to as such, and never as "Beatles". In these cases, the group considers the initial article significant, and it will be the access point consulted by many users. An extreme example is the group "The The", which would look absurd with the initial "The" omitted or inverted.'

There are good reasons for sorting on 'The', says Browne, and good reasons for ignoring it. She suggests listing 'The' items twice: under 'The' *and* under the second word in the entry. Lest that create unmanageably long lists of entries starting with 'The', she offers other alternatives.

What should indexers do about entries starting with 'The'?	Double entry with and without *The*	**Win-win situation proposed by Glenda Browne, the.**
Indexers apply generally accepted rules of indexing and also use individual judgement, always thinking: 'Where would a user look for this item?'. Since some users do not know the rules, I suggest we should put entries starting with 'The' wherever those users might look; that is, both sorted on 'The' and as a double entry under the second word. Where the length or complexity of the index is a factor, then a cross-reference would replace the double entry, and the indexer would have to decide which form to prefer.	There are good reasons for sorting on 'The', and good reasons for ignoring it. The win-win situation is therefore to have double entry of titles, place names and corporate names under 'The' *and* under the second word in the entry. If this creates unmanageably long lists of entries starting with 'The', then a reference could be added. For example: *The ..., to search for titles and names starting with 'The', see the second word in the entry.* This quickly lets users know what rule you are using. Alternatively, information about the way you have dealt with 'The' can be included in the introduction to the index.	

Internationally, the 'The' problem is not *the* problem – it is merely *a* problem. Browne makes this clear at the very start of her paper, with a quotation from indexing maven Hans H. Wellisch: 'Happy is the lot of an indexer of Latin, the Slavic languages, Chinese, Japanese, and some other tongues which do not have articles, whether definite or indefinite, initial or otherwise.'

For her study of the 'The' problem, Glenda Browne was awarded the 2007 Ig Nobel Prize in literature.

Browne, Glenda (2001). 'The Definite Article: Acknowledging "The" in Index Entries.' *The Indexer* 22 (3): 119–22.

TO WRESTLE A BAD METAPHOR

Carl Phillips, Brian Guenzel, and Paul Bergen are hopping mad about bad metaphors. Writing in the almost-poetically-named *Harm Reduction Journal*, they pull on their metaphorical boxing gloves. Stepping into the ring, so to speak, they issue a forthright statement: 'Anti-harm-reduction advocates sometimes resort to pseudo-analogies to ridicule harm reduction. Those opposed to the use of smokeless tobacco as an alternative to smoking sometimes suggest that the substitution would be like jumping from a 3-story building rather than 10-story, or like shooting yourself in the foot rather than the head.' Following their summary of these two disagreeable metaphors, Phillips, Guenzel, and Bergen then proceeded to administer a good thrashing.

They collected several versions of the 'jump from a building' metaphor. Some metaphor makers, they report, have 'likened smoking to falls from at least the 10th floor and smokeless tobacco to falls from at least the 3rd; we found numbers as high as 50 and 30'. These are beneath contempt, they explain, because 'anyone with passing familiarity with the human body and Earth's gravity should be aware that falls from the 10th story are almost always fatal'.

Perhaps unsure of their own familiarity with the human body and Earth's gravity, they conducted a review of the available literature on mortality rates as a function of free-fall distance.

'It is surprising', they write, 'how little information is published on the topic ... The literature suggests that falls from up to the 3rd story are most always survived, with the death rate increasing sharply and approaching 100% over the next three or four stories ... More accurate analogies might actually be fairly useful in painting the picture for consumers. A nontrivial portion of young men have probably jumped from a 2nd story window, but few would dare jump from the 4th.'

This is their main line of attack. They go at it from other directions, too.

In the second bout, they pummel the 'gunshot to the head' metaphor, applying a devastating one-two punch:

1) 'It is immediately obvious that the gunshot metaphor is absurd: If someone was faced with the choice of shooting himself in the head and shooting himself in the foot or leg, the latter option is quite obviously better from a health outcomes perspective.'

2) 'Mortality risk from self-inflicted gunshot wounds to the head dwarfs that from smoking, while foot wounds, though they have a low mortality rate, have a high probability of permanent debilitating orthopedic damage, a risk absent in tobacco use.'

Phillips and Bergen are at the University of Alberta in Edmonton. Guenzel is at the Center for Philosophy, Health, and Policy Sciences in Houston, Texas.

In the end, they let slip what's most got up their nose. 'The metaphors', they write, 'exhibit a flippant tone that seems inappropriate for a serious discussion of health science.'

Phillips, Carl V., Brian Guenzel, and Paul Bergen (2006). 'Deconstructing Anti-harm-reduction Metaphors: Mortality Risk from Falls and Other Traumatic Injuries Compared to Smokeless Tobacco Use.' *Harm Reduction Journal* 3: 15.

IN BRIEF

'THE CASE OF THE BURLY WEE MAN'
(published in the *Archives of Environmental Health*, 1974)

COLONOSCOPY SOCIOLOGICA

The passing years make it difficult to remember how much excitement arose when Sue Ziebland and Catherine Pope published their epic report 'The Use of the Colon in Titles of British Medical Sociology

Conference Papers, 1970 to 1993'.

Ziebland was then at the department of public health medicine of the Camden and Islington health authority in London. She has since passed through the digestive system of academia and emerged at the University of Oxford. Pope was at the London School of Hygiene and Tropical Medicine. Time and circumstance have deposited her at the University of Bristol.

Together Ziebland and Pope explored the colons of several British researchers. Their report appeared in the *Annals of Improbable Research*. It shed light on a problem that distressed many social scientists: how to properly devise the titles of their conference papers. This is how Ziebland and Pope described it: 'Unless the presenter is of an un-usually retiring disposition, there will be a desire to choose a punchy, attention-grabbing title in the hope of attracting a large and lively audience. However, sooner or later the truth will out, and it is clearly in one's interest to make some mention of the actual subject of the work. The favoured solution is to use a colon, separating the snappy from the prosaically descriptive, as in: "Sex and Drugs: Women's Use of Aspirin".'

Ziebland and Pope examined trends in the use of the colon in paper titles using evidence from a particular annual conference. They considered every paper that was listed in the printed programmes from the first annual conference in 1969, up to the 1993 meeting.

Their analysis is based on the percentage of the total number of papers per year that include one or more colons in the title. They tallied each paper as a single occurrence, even a 1979 paper that in-cluded five colons. (Alas, they do not tell us the name of that paper.)

They discovered that the percentage of paper titles increased almost continuously during the 1970s and 1980s. From the mid-1980s onwards, a steady forty to forty-eight percent of titles included a colon. In the year 1985, a staggering fifty-seven percent featured colons. This anomaly, Ziebland and Pope wrote, 'has no obvious explanation'.

The colon has fascinated scholars for generations. More than a decade before Ziebland and Pope's colonic examination, the noted and footnoted scholar J. T. Dillon, of the University of California, Riverdale, performed three historical endoscopies of the academic colon. They are:

The Emergence of the Colon: An Empirical Correlate of Scholarship (*American Psychologist*, 1981)
Functions of the Colon: An Empirical Test of Scholarly Character (*Educational Research Quarterly*, 1981)
In Pursuit of the Colon: A Century of Scholarly Progress:
   1880–1980 (*Journal of Higher Education*, 1982)

Incisive and exciting as these studies may have been when published, they are now seen as period pieces.

HIGHLIGHTS OF HIGHLIGHTING

The practice of reading textbooks for pleasure is just as lively now as
it has ever been. More people buy textbooks – actually spend their
own money to do it – now than ever before. And in deciding what to
buy, they (or should I say, 'we') are kids in a candy shop. There is an
ever-growing number of specialized subjects for which textbooks exist, and so the variety of textbooks on offer is always increasing. Even
if you somehow manage to exhaust the cream of one genre, you can
easily find another to sample.

An untimid reader can find lots of good, meaty reads packed with
literary merit. Like the best novels, many of the textbooks in forestry
management, ergodic theory, multinational auditing, and thousands
of other genres try to fill a reader's mind with ideas and words that,
at first read, really do feel completely novel.

But that's not the best part. Used textbooks offer one thing more
to beguile the leisure-time reader.

For many of us, the highlight of reading used textbooks is the
highlighting, the lines previous readers have drawn under, around,
or through particular words or passages. Good highlighting makes
any used textbook worth the purchase. Bad highlighting makes it
even better. And in buying highlighted textbooks, you sometimes get
a double bonus: despite the carefully added interest, they often have
drastically reduced price tags.

Of course, not everyone's pulse races at the sight of a textbook. H.
G. Wells was outspoken about this. In 1914, he put textbooks in their
supposed place, which to him was fifth in a list of derogatory words
he used to describe bad education: 'thin, ragged, forced, crammy,

text-bookish, superficial'. Wells, for all his insights into science, into humanity, into the future, into the etc., was somehow not seeing the good parts – not even the highlights! – of textbooks.

Vicki Silvers and David Kreiner, of Central Missouri State University, stepped on to the scene eighty-three years later, with a study called 'The Effects of Pre-Existing Inappropriate Highlighting on Reading Comprehension'. 'Textbook highlighting is a common study strategy among college students', Silvers and Kreiner wrote, using the academese that their profession demands. Then they described their experiments.

First, they had students read a passage of text. Some students had text that was highlighted appropriately. Some had text that was highlighted inappropriately. Others had Spartan, unhighlighted text. Silvers and Kreiner then tested how well the students comprehended the text. Those with the inappropriate highlighting scored much lower than the others. A second experiment showed that even when students were warned about the inappropriate highlighting, they had trouble ignoring it.

In 2002, Silvers and Kreiner were awarded the Ig Nobel Prize in literature. At the awards ceremony, they offered one piece of advice: 'Don't buy a textbook that was highlighted by an idiot.' I'm not sure I'd agree.

Silvers, Vicki, and David Kreiner (1997). 'The Effects of Pre-Existing Inappropriate Highlighting on Reading Comprehension.' *Reading Research and Instruction* 36: 217–23.

HEY, DUDE...

The story of dude – its rise, its role, its rich history as a word – takes twenty-five pages to tell. University of Pittsburgh linguistics professor Scott Fabius Kiesling's analysis of dude carries the title 'Dude'. It occupies a stylish chunk of the autumn 2004 issue of the journal *American Speech*.

Kiesling's tale accords with the pithy history of dude that you'll find in the Oxford English Dictionary. Of American origin, dude in the 1880s was 'a name given in ridicule to a man affecting an exaggerated fastidiousness in dress, speech, and deportment'. A few decades later, a dude was 'a non-westerner or city-dweller who tours or stays in the west of the US, especially one who spends his holidays on a ranch; a tenderfoot'. Nowadays, a dude is the object of more-than-just-self-esteem. Today's dude is 'any man who catches the attention in some way; a fellow or chap, a guy. Hence also approvingly, especially applied to a member of one's own circle or group.'

Kiesling delves deep into the modern dude, the dude of whom we hear speak wherever young Americans roam. He provides context for those whom the world may have passed by. 'Older adults', he writes, 'baffled by the new forms of language that regularly appear in youth cultures, frequently characterize young people's language as "inarticulate", and then provide examples that illustrate the specific forms of linguistic mayhem performed by "young people nowadays".'

He then gets down to business, outlining 'the patterns of use for dude, and its functions and meanings in interaction'. Dude, we learn, is: (a) used mostly by young men to address other young men; (b) a general address term for a group (same or mixed gender); and (c) a discourse marker that generally encodes the speaker's stance to his or her current addressee(s). Best of all, 'Dude indexes a stance of cool solidarity, a stance which is especially valuable for young men as they navigate cultural Discourses of young masculinity.'

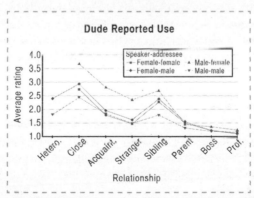

Note from study: The label 'Hetero' refers to 'heterosexual intimate relationships', and though 'there were responses for male-male and female-female categories ... it is clear from the students who gathered the data that not all respondents understood the intimate nature of this category'

Kiesling attributes the sudden blossoming of dude, in the 1980s, to the actor Sean Penn, who played the role of Jeff Spicoli in the film *Fast Times at Ridgemont High*. Penn, in his Spicoli persona, is 'the do-nothing, class-cutting stoned surfer' who takes 'a laid-back stance to the world, even if the world proves to be quite remarkable'. Kiesling confides that he was a teenager at the time the film came out, and that 'many young men glorified Spicoli, especially his nonchalant blindness to authority and hierarchical division'.

The bulk of 'Dude' is technical, an exploration of data gathered by students in Pittsburgh. Each student wrote down the first twenty usages of the word 'dude' they heard during a three-day period. These Kiesling compiled into what he calls 'The Dude Corpus'. The corpus awaits the scrutiny of future dudes and scholars of dude, who may see in it things that are invisible to us.

Kiesling, Scott F. (2004). 'Dude.' *American Speech* 79 (3): 281–305.

HAZARDS OF BOBBING

New parents beware! is the implied theme of a new study called 'Who Do You Look Like? Evidence for the Existence of Facial Stereotypes for Male Names'.

The researchers begin with this little shocker: 'Choosing a name for a forthcoming baby occupies a good deal of time for most expectant parents ... Few worry about whether the name will provoke a facial stereotype in the minds of others (hmm ... he doesn't look like a "Bob"), but, as the present research suggests, this may be yet another potential worry to have when one selects a name for one's progeny.'

Scientists Melissa A. Lea, Robin D. Thomas, Nathan A. Lamkin, and Aaron Bell hammer home the unfairness of the situation. 'This is an especially provocative suggestion', they write, 'as names are usually chosen before or immediately after birth, certainly before any knowledge becomes available of what the child may look like when they are adults.'

All of the team members are associated with Miami University of Ohio. The university issued a press release that announced 'researchers at Miami University think they know why you can remember some peoples' names but not others'. They've shown quantitatively that certain names are associated with certain facial features. For example, when people hear the name "Bob" they have in mind a larger, round face than when they hear a name such as "Tim" or "Andy".'

The study includes a pair of photos — on the left, a tousle-haired young man in a white shirt; on the right a bald, droopy-eyed fellow wearing what might be a striped prison outfit. The researchers say, when shown these two pictures, 'audience members overwhelmingly agree that the man on the left is named "Tim" and the man on the right is named "Bob".'

Figure: 'Audience members overwhelmingly agree that the man on the left is named "Tim" and the man on the right is named "Bob".'

This, however, was not what happened in the experiment.

In the experiment, software was used to create idealized, hairless faces for each of fifteen names: Andy, Brian, Joe, Justin, Rick, Bill, Dan, John, Mark, Tim, Bob, Jason, Josh, Matt, and Tom. Then a group of volunteers was given printed cards – each card had one of those pictures or one of those names – and told to match up pictures and names.

Most agreed that Bobs are Bob-like. Many agreed that the Bills are Bill-like and that the Toms are Tom-like. There was not so much agreement as to the other faces and names.

Of course, these research results are definitive only for those particular faces and those particular names, and only as they struck one particular group of student volunteers on one particular day.

The researchers say they were inspired, at least a little, by a study done nearly a century ago. In 1916, a researcher based at Cornell University, in Ithaca, New York, wanted to understand what she called 'the nature of the psychological response to proper names of unknown persons'. Basically, she asked: What sort of person is named Rupzóiyat?

In particular, the researcher, who herself was named G. English, wanted to test a theory proposed by a Swiss psychologist named Édouard Claparède. The theory says that, 'other things equal, names consisting of heavy or repeated syllables call forth images of fat, heavy-set, bloated, or slightly ridiculous individuals; a short and sonorous name, on the other hand, suggests slender and active persons, etc.'.

English concocted fifty 'nonsense names' – names stuck together with syllables she chose at random. Then she tested the names on eight people. Here's how she described the experiment: 'Each name was pronounced three times over, the experimenter being careful to pronounce it slowly, distinctly, and (as nearly as possible) always in the same manner. [Then the observer was asked] to describe the person that "must belong to the name"'.

English's fifty names: Chérin, Póisher, Kilom, Koikert, Vázal, Dáwfisp, Zóque, Spren, Dáwthô, Rupzóiyat, Blag, Lísrix, Thaspkûwhin, Kîrd'faumish, Génras, Tháchô, Brob, Zóitû, Kóldak, Múrbix, Chermt-gáwkonv, Bóppum, Vúshap, Grib, Watshóiquol, Móiki, Hoxzáuwhuk, Gáwthû, Zé'the, Gówsü, Déznep, Wîtaw, Thôbonf, Mávquawpûnt, Stisk, Tówbant, Táquû, Skamth, Quajnûmeth, Bünoy, Drup, Gúklal, Pófmoj, Spux, Jíkzel, Snemth, Thúbtawkarnth, Línrêwex, Gronch, and Túpjoz.

English also asked the observers to try to spell the names back to her. She didn't care whether they got the spelling right. She just wanted to ensure that they had heard her clearly.

The results disappointed her: 'In only five cases was there anything like agreement among all observers as to sex or other characteristics. Rupzóiyat was reported as a young man by all observers; Bóppum was said to be a tall, fat or large man by six observers.' Of the eight observers, 'five thought Zé'the must be a girl; six reported Grib as a small man; and five reported Kîrd'faumish as a strong or big man. For all the remainder there was disagreement.'

English decided that 'there is no constant or uniform tendency among these observers [to] imagine a similar type of individual for the same name.'

> XXXI. ON THE PSYCHOLOGICAL RESPONSE TO
>   UNKNOWN PROPER NAMES
>
> By G. ENGLISH

She mused about the way Charles Dickens played with nonsense names. But she concluded that maybe Dickens – and maybe all of us

– only occasionally see a person's name as some sort of guide to their nature: 'We know that Dickens came to [evolve the name] Chuzzlewit through Sweezleden, Sweezleback, Sweezlewag, Chuzzletoe, Chuzzleboy, Chubblewig, and Chuzzlewig. The name was significant to him; and yet there were various types of Chuzzlewit, as there were various types of Nickleby. Indeed, the applicability of a surname to all the members of a family must, one would suppose, tend to prevent our attaching any special import to the name's physiognomy.'

Lea, Melissa A., Robin D. Thomas, Nathan A. Lamkin, and Aaron Bell (2007). 'Who Do You Look Like? Evidence for the Existence of Facial Stereotypes for Male Names.' *Psychonomic Bulletin & Review* 14 (5): 901–7.
English, G. (1916). 'On the Psychological Response to Unknown Proper Names.' *American Journal of Psychology* 27: 430–34.

CALL FOR INVESTIGATORS

The Rhyming Monikers Research Citation Bibliography Project, announced here, is searching for additions to its collection of outstanding research on which the co-authors' names rhyme.

Definition: For the purposes of the project, a Rhyming Moniker involves sound correspondence in at least the terminal syllable of the co-authors' surnames. The first specimen in the collection, unearthed by Investigator Russell Mortishire-Smith, serves as a model organism:

'Measurement of Long-Range 13C-13C J Couplings in a 20-kDa Protein-Peptide Complex' by Ad Bax, David Max, and David Zax (published in the *Journal of the American Chemical Society*, 1992). The authors are at what is abbreviatingly referred to as the Lab. Chem. Phys., Natl. Inst. Diabetes Dig. Kidney Dis., Bethesda, Maryland.

Bax, Max, Zax. The names ring out. They sing out. An impressive contribution.

Purpose: The collection will be made available to other researchers for cross-variable and meta-analysis.

If you have identified a new specimen, please send to marca@improbable.com with the subject line:
RHYMING MONIKERS RESEARCH CITATION COLLECTION

TROLLING FOR ANNOYANCE

Trolls – call them Internet trolls, if you like – are distant behavioural kin to *Plasmodium falciparum*, a protozoan parasite that causes malaria in large numbers of human beings. Both kinds of parasite are maddeningly difficult to suppress. They manage, again and again, to return after we thought we'd seen the last of them. Each can, if left untreated, cause agony or worse.

These trolls infect any place where people gather electronically to converse by writing comments to each other. Trolls creep into and crop up anywhere they can, wheedling for attention in chat rooms, listservs, twitter streams, blogs, and as you may have noticed, in the comments section of online news articles.

One of the many annoying things about Internet trolls is that it's difficult to define precisely, with academic rigour, what they do. Claire Hardaker, a lecturer at University of Central Lancashire's department of linguistics and English language, took up the challenge. Her study called 'Trolling in Asynchronous Computer-Mediated Communication' is published somewhat counter-intuitively in the *Journal of Politeness Research*.

Hardaker presented an early form of the paper to a mostly troll-free audience at the Linguistic Impoliteness and Rudeness conference held at her university in 2009.

After much research and hard work, Hardaker came up with a working definition. A troll is someone 'who constructs the identity of sincerely wishing to be part of the group in question, including professing, or conveying pseudo-sincere intentions, but whose real intention(s) is/are to cause disruption and/or to trigger or exacerbate conflict for the purposes of their own amusement'.

Example (6)	[060318]
7. A	If you are a troll ... I'm sure you'd never admit it. If there even is a *real* pony in all of this, I feel very sorry for it.
8. B	I am not a troll ... I am only doing what my vet has advised me to do.

Detail: Trolling for an annoying example

She arrived at this after much trolling (in a very different sense of that word) through data. Lots of data. A '172-million-word corpus of unmoderated, asynchronous computer-mediated communication', a nine-year collection of commentary in an online discussion group about horseback riding. She focused in on the huge number of passages where people mentioned trolls, trolling, trolled, trollish, trolldom, and other variations on the key word 'troll'.

Distilling the wisdom of the horse-talk crowd, Hardaker set up this handy guide to interacting with trolls: 'Trolling can (1) be *frustrated* if users correctly interpret an intent to troll, but are not provoked into responding, (2) be *thwarted*, if users correctly interpret an intent to troll, but counter in such a way as to curtail or neutralize the success of the troller, (3) *fail*, if users do not correctly interpret an intent to troll and are not provoked by the troller, or, (4) *succeed*, if users are deceived into believing the troller's pseudo-intention(s), and are provoked into responding sincerely. Finally, users can *mock troll*. That is, they may undertake what appears to be trolling with the aim of enhancing or increasing affect, or group cohesion.'

Any comments?

Hardaker, Claire (2010). "Trolling in Asynchronous Computer-Mediated Communication: From User Discussions To Academic Definitions.' *Journal of Politeness Research* 6 (2): 215–42.

IN BRIEF

'CONSEQUENCES OF ERUDITE VERNACULAR UTILIZED IRRESPECTIVE OF NECESSITY: PROBLEMS WITH USING LONG WORDS NEEDLESSLY'

by Daniel M. Oppenheimer (published in *Applied Cognitive Psychology* and honoured with the 2006 Ig Nobel Prize in literature)

THE PSYCHOLOGY OF REPETITIVE READING

A typical adult knows almost nothing about the psychology of repetitive reading. This is not surprising. Research psychologists, as a group, know little about the subject, though some have attempted to close the gap.

Human beings can be induced to read repetitively. In one experiment, a scientist named N. Borgovsky asked two hundred subjects to read a repetitive essay. The essay consisted of a single paragraph repeated several times. Each subject was told beforehand that the essay was highly repetitive. The result was surprising. Ninety-two percent of the subjects read the essay completely from beginning to end.

Borgovsky began his experiment by recruiting several dozen people, whom he asked to be his research subjects. A typical adult knows almost nothing about the psychology of repetitive reading. (This is not surprising. Research psychologists, as a group, know little about the subject, though some have attempted to close the gap.) So Borgovsky sat his subjects down in a room, and explained that human beings can be induced to read repetitively. In one experiment, he told them, a scientist asked two hundred subjects to read a repetitive essay. The essay consisted of a single paragraph repeated several times. Each subject was told beforehand that the essay was highly repetitive. The result was surprising. Ninety-two percent of the subjects read the essay completely from beginning to end.

After giving his subjects that background information, Borgovsky described his own experiment in great detail. The experiment was based on a book he had read. The book was based on the idea that human beings can be induced to read repetitively. In one experiment, a scientist asked two hundred subjects to read a repetitive essay. The essay consisted of a single paragraph repeated several times. Each subject was told beforehand that the essay was highly repetitive. The result was surprising. Ninety-two percent of the subjects read the essay completely from beginning to end.

After Borgovsky carried out his experiment, he published a report. Called 'The Psychology of Repetitive Reading', it explains that human beings can be induced to read repetitively. In one experiment, a scientist – Borgovsky, in fact – asked two hundred subjects to read a repetitive essay. The essay consisted of a single paragraph repeated several times. Each subject was told beforehand that the essay was highly repetitive. The result was surprising. Ninety-two percent of the subjects read the essay completely from beginning to end.

MEANING? MEANING? MEANING?

Yes, yes, yes – there are many ways to repeat yourself. Some are more meaningful than others, says a clever linguist in the Netherlands.

Technically speaking, 'Yes, yes, yes!' is an example of 'multiple sayings in social interaction'. Tanya Stivers has pursued, bagged, and intensively studied a small herd of multiple sayings. Her thirty-three-page report about them, called '"No no no" and Other Types of Multiple Sayings in Social Interaction', was published in the journal *Human Communication Research*.

Stivers is based at the Max Planck Institute for Psycholinguistics, in Nijmegen, The Netherlands. She decided to look at just one species in the multiple-sayings menagerie. The repetition 'Okay okay okay' interests Stivers a lot. The repetition 'Okay. Okay. Okay' does not.

'Okay okay okay' is 'a single stretch of talk'.

'Okay. Okay. Okay', on the other hand, is 'multiple packages'.

She believes that the one is, in a way, quite different from the other. Repeating little phrases as a single lump can imply something deep and simple: that the other person's whole course of action is problematic and should be halted.

Stivers illustrates her point with lots of snatches of conversation. Her report delves into the technical aspects of certain uses of 'Yes yes yes', 'No no no', 'Right right right', 'I'll eat 'em / I'll eat 'em / I'll eat 'em', 'a'right / a'right / a'right', and 'I see / I see / I see'.

Most of her examples come from English-language conversations, but Stivers says that 'the practice has been found in Catalan, French, Hebrew, Japanese, Korean, Lao, and Russian, as well'.

Tanya Stivers is part of a small gaggle of scholars who call themselves conversation analytic researchers. Conversation analytic researchers study the anatomy and physiology of people's jabberings. They record people talking, then have someone transcribe the recordings. Then they analyse / analyse / analyse.

Conversation is complicated stuff, despite the ease with which people yak / yak / yak together. It can be difficult for outsiders – people who are not conversation analytic researchers – to appreciate that these professionals need some unusual skills.

A taste of professionally boiled, sliced conversational analysis can seem off-putting to the casual conversationalist. Here is a not-unusual sample, written by Thomas Holtgraves, of Ball State University in Muncie, Indiana: 'Conversation analytic researchers have demonstrated that conversationalists do appear to be sensitive to the occurrence of dispreferred markers'.

Such technical lingo can make it difficult for people who are not conversation analytic researchers to see what people who are conversation analytic researchers are talking about. This is sad, because what conversation analytic researchers talk about, mostly, is the conversations of people who are not conversation analytic researchers. And the researchers rejoice, because their research tells them that repetition is not – repeat, *not* – boring.

Stivers, Tanya (2004). '"No no no" and Other Types of Multiple Sayings in Social Interaction.' *Human Communication Research* 30 (2): 260–93.
Holtgraves, Thomas (2000). 'Preference Organization and Reply Comprehension.' *Discourse Processes* 30 (2): 87–106.

CALL FOR SUBMISSIONS

If you know of any improbable research – the sort that makes you laugh and think, and that you think will make other people laugh, too – I would be delighted and grateful to hear about it.

Please email me at marca@improbable.com with an improbable subject line of your choosing.

ACKNOWLEDGEMENTS

Thank you to the good editors (and in my experience, they are exactly that) who got and kept me writing for the *Guardian* and whose suggestions and advice and criticism have almost always been 'spot on' (as British people say in books) and encouraging. In order of appearance, they are Tim Radford, Will Woodward, Claire Phipps, Donald MacLeod, and Alice Wooley.

Thank you to my wife, Robin. Thanks to my parents for endowing me with tolerance and curiosity for random assortments. Thanks to my many colleagues and friends and readers (those categories blend) who together comprise Improbable Research – both the uppercase version and the larger, lowercase version – and the Ig Nobel gang. Look on the web site, www.improbable.com, and you'll see many of their names and much of their influence.

Thanks to the three people most responsible for conjuring whatever it was they conjured that caused this book to apparate: Regula Noetzli and Caspian Dennis, ace agents; and Robin Dennis, ace editor. These Dennises, I believe, are not related to each other, except in the world of books.

Particular thanks to each of the many people who told me about things that wound up in this book. They have been kind in sharing their discoveries with me, that I might share them with you. Here are some of them: Claudio Angelo, Catherine L. Bartlett, Michael L. Begeman, John Bell, Charles Bergquist, Lisa Birk, John D. Bullough, Peter Carboni, the Chemical Heritage Foundation, Francesca Collins, Lauradel Collins, Jim Cowdery, Fuzz Crompton, Missy Cummings, Wim Crusio, Kristine Danowski, David Derbyshire, Betsy Devine, Paola Devoto, Tatiana Divens, Matthias Ehrgott, Stanley Eigen, Steve Farrar, Rose Fox, Stefanie Friedhoff, Andrea Gaddini, Martin Gardiner, Rebecca German, David Gevirtz, Tom Gill, Max Glaskin, Diego Golombek, N. Hammond, Ron Hassner, Mark Henderson, Simon

Hudson, Alok Jha, Torbjörn Karfunkel, Mark Keiser, David Kessler, Erwin Kompanje, Scott Langill, Tom Lehrer, T. Leighton, Jill LePore, Alan Litsky, Julia Lunetta, Donald MacLeod, James Mahoney, William J. Maloney, G. N. Martin, Neil Martin, Les Martinsson, Maryn McKenna, Chris McManus, Fernando Merino, Rosie Mestel, Katherine Meusey, Kees Moeliker, Jean Monahan, Harold Morowitz, Gabriel Nève, Scott A. Norman, Charles Oppenheim, Eduardo B. Ottoni, Rich Palmer, Ruth Parrish, Michael Ploskonka, Bella Plouffe, Stavros Poulos, Hanne Poulsen, Gus Rancatore, James Randerson, Thomas A. Reisner, R. Roberts, Geneva Robertson, Ian Sample, Reto Schneider, M. Schreiber, Sally Shelton, Adrian Smith, Annette Smith, Andrew N. Stephens, Geri Sullivan, Frank Sutman, B. E. Swetman, Vaughn Tan, Tony Taylor, Mary Thomson, Richard Wassersug, Corky White, Amity Wilczek, Michael Wolfson, and Jan Wooten.

EXTRA CITATIONS

IN BRIEF

Al Fallouji, M. (1990). 'Traumatic Love Bites.' *British Journal of Surgery* 77: 100–1.

Buchanan, D. R., D. Lamb, and A. Seaton (1981). 'Punk Rocker's Lung: Pulmonary Fibrosis in a Drug Snorting Fire-Eater.' *British Medical Journal* 283: 1661.

Cassaro, A., and M. Daliana (1992). 'Impaction of an Ingested Table Fork in a Patient with a Surgically Restricted Stomach.' *New York State Journal of Medicine* 92 (3): 115.

Cheng, G., Z. Xuand, and J. Xu (2005). 'Vision of Integrated Happiness Accounting System in China.' *Acta Geographica Sinica* 60 (6): 893–901.

Coolidge, Frederick L. (1999). 'My Grandmother's Personality: A Posthumous Evaluation.' *Journal of Clinical Geropsychology* 5 (3): 215–19.

Earles, C. M., A. Morales, and W. L. Marshall (1988). 'Penile Sufficiency: An Operational Definition.' *Journal of Urology* 139 (3): 536–38.

Foley, J., and J. J. Sheuring (1966). 'Cause of Microbial Death during Freezing in a Soft Serve Ice Cream Freezer.' *Journal of Dairy Science* 49 (8) 928–32.

Gordon, Christopher J., and Elizabeth C. White (1982). 'Distinction Between Heating Rate and Total Heat Absorption in the Microwave-Exposed Mouse.' *Physiological Zoology* 55 (3): 300–8.

Krause, D., D. Ick, and H. Treu (1981). 'Successful Insemination Experiments with Cryopreserved Sperm from Wild Boars.' *Zuchthygiene*.

Liu, Bing, Liang-Ping Ku, and Wynne Hsu (1997). 'Discovering Interesting Holes in Data.' *Proceedings of Fifteenth International Joint Conference on Artificial Intelligence*, Nagoya, Japan: 930–35.

McVeigh, Brian J. (2000). 'How Hello Kitty Commodifies the Cute, Cool, and Camp: "Consumutopia" Versus "Control" in Japan.' *Journal of Material Culture* 5 (2): 225–45.

Oppenheimer, Daniel M. (2006). 'Consequences of Erudite Vernacular Utilized Irrespective of Necessity: Problems with Using Long Words Needlessly.' *Applied Cognitive Psychology* 20: 139–56.

Ortmann, C., and A. DuChesne (1998). 'A Partially Mummified Corpse with Pink Teeth and Pink Nails.' *International Journal of Legal Medicine* 111: 35–37.

Pories, Walter J. (2001). 'The Cow with Zits.' *Current Surgery* 58 (1): 1.

Seiffert, H. J. (2000). 'A Cute Characterization of Acute Triangles.' *American Mathematical Monthly* 107 (5): 464.

Smith, Geoff P. (1995). 'How High Can a Dead Cat Bounce?: Metaphor and the Hong Kong Stock Market.' *Linguistics and Language Teaching* 18: 43–57.

Smith, Thomas J., Bruce E. Hillner, and Harry D. Bear (2003). 'Taking Action on the Volume-Quality Relationship: How Long Can We Hide Our Heads in the Colostomy Bag?' *Journal of the National Cancer Institute* 95 (10): 695–97.

Spears, A. S. (1994). 'Attempted Suicide or Hitting the Nail on the Head: Case Report.' *Journal of the Florida Medical Association* 81 (12): 822–23.

Stitt, W. Z., and A. Goldsmith (1995). 'Scratch and Sniff: The Dynamic Duo.' *Archives of Dermatology* 131: 997–99.

Thompson, D. G. (2001). 'Descartes and the Gut: "I'm Pink Therefore I Am".' *Gut*, 2001) 49: 165–66.
Witts. Richard (2005). 'I'm Waiting for the Band: Protraction and Provocation at Rock Concerts.' *Popular Music* 24 (1): 147–52.
N.A. (1974). 'The Case of the Burly Wee Man' (published in the *Archives of Environmental Health* 28 (5): 297–98.

MAY WE RECOMMEND

Anders Barheim and Hogne Sandvik (1994). 'Effect of Ale, Garlic, and Soured Cream on the Appetite of Leeches.' *BMJ* 209: 1689.
Bollinger, S. A., S. Ross, L. Oesterhelweg, M. J. Thali, and B. P. Kneubeuhl (2009). 'Are Full or Empty Beer Bottles Sturdier and Does Their Fracture-Threshold Suffice to Break the Human Skull?' *Journal of Forensic and Legal Medicine* 16: 138–42.
Bolt, Michael, and Daniel C. Isaksen (2010). 'Dogs Don't Need Calculus.' *College Mathematics Journal* 41 (1): 10–16.
Bubier, Norma E., Charles G. M. Paxton, P. Bowers, D. C. Deeming (1998). 'Courtship Behaviour of Ostriches (*Struthio camelus*) Towards Humans Under Farming Conditions in Britain.' *British Poultry Science* 39 (4): 477–81.
Coventry, K. R., and B. Constable (1999). 'Physiological Arousal and Sensation-Seeking in Female Fruit Machine Gamblers.' *Addiction* 94 (3): 425–30.
Griffin, John M., and Jin Xu (2009). 'How Smart Are the Smart Guys? A Unique View from Hedge Fund Stock Holdings.' *Review of Financial Studies* 22 (7): 2531–70.
Griffin, Michael J., and R. A. Hayward (1994). 'Effects of Horizontal Whole-Body Vibration on Reading.' *Applied Ergonomics* 25 (3): 165–69.
Hopton, Robert, Steph Jinks, and Tom Glossop (2010). 'Determining the Smallest Migratory Bird Native to Britain Able to Carry a Coconut.' *Journal of Physics Special Topics* 9 (1).
Krippner, Stan, Monte Ullman, and Bob Van de Castle (1973). 'An Experiment in Dream Telepathy with "The Grateful Dead".' *Journal of the American Society of Psychosomatic Dentistry and Medicine* 20: 9–17.
Miller, Geoffrey, Joshua M. Tybur, and Brent Jordan (2007). 'Ovulatory Cycle Effects on Tip Earnings by Lap Dancers: Economic Evidence for Human Estrus?' *Evolution and Human Behavior* 28 (6): 375–81.
Nirapathpongporn, Apichart, Douglas H. Huber, and John N. Krieger (1990). 'No-Scalpel Vasectomy at the King's Birthday Vasectomy Festival.' *Lancet* 335: 894–95.
Pennings, Timothy J. (2010). 'Do Dogs Know Calculus?' *College Mathematics Journal* 34 (3): 178–182.
Randall, Brad (2009). 'Blood and Tissue Spatter Associated with Chainsaw Dismemberment.' *Journal of Forensic Sciences* 54 (6): 1310–14.
Stephens, Richard, John Atkins, and Andrew Kingston (2009). 'Swearing as a Response to Pain.' *Neuroreport* 20: 1056–60.
Traub, Stephen J., Robert S. Hoffman, and Lewis S. Nelson (2001). 'Pharyngeal Irritation After Eating Cooked Tarantula.' *Internet Journal of Medical Toxicology* 39: 562.
US National Institute of Health (1993). 'Fatalities Attributed to Entering Manure Waste Pits.' *Morbidity and Mortality Weekly Report* 42 (17): 325–29.
Willan, Derek, ed. (1994). *Greek Rural Postmen and Their Cancellation Numbers*. N.P.: Hellenic Philatelic Society of Great Britain.

AN IMPROBABLE INNOVATION

Bodnar, Elena N., Raphael C. Lee, and Sandra Marijan (2007). 'Garment Device Convertible to One or More Facemasks.' US Patent no. 7,255,627, 14 August.

Imai, Makoto, Naoki Urushihata, Hideki Tanemura, Yukinobu Tajima, Hideaki Goto, Koichiro Mizoguchi, and Junichi Murakami (2009). 'Odor Generation Alarm and Method for Informing Unusual Situation.' US Patent application no. 2010/0308995 A1, 9 December.

Keogh, John (2001). 'Circular Transportation Facilitation Device.' Australian Innovation Patent no. 2001100012, 2 August.

Li, Zhengcai (2007). 'A Tittle Obliquity Measurer.' International Patent application no. PCT/CN2007/003282, 6 December.

Miller, Gregg A. (1999). 'Surgical Method and Apparatus for Implantation of a Testicular Device.' US Patent no. 5,868,140, 9 February.

Nutting, William B. (1971). 'Kiss Throwing Doll.' US Patent no. 3,603,029, 7 September.

Scruggs, Donald E. (2011). 'Edged Non-horizontal Burial Containers.' US Patent no. 8,046,883, 1 November.

Smith, Frank J., and Donald J. Smith (1977). 'Method of Concealing Partial Baldness.' US Patent no. 4,022,227, 10 May.

CALL FOR INVESTIGATORS

Bax, Ad, David Max, and David Zax (1992). 'Measurement of Long-Range 13C-13C J Couplings in a 20-kDa Protein-Peptide Complex.' *Journal of the American Chemical Society* 114 (17): 6923–25.

ILLUSTRATION CREDITS

Grateful acknowledgement is made to the many researchers and inventors whose work is illustrated in these pages, sometimes with illustrations, sometimes without.

Selected steps from 'Method of Concealing Partial Baldness' (p. 20) from US Patent no. 4,022,227
From 'Garment Device Convertible to One or More Facemasks' (p. 30) from US Patent no. 7,255,627
Height vs. distance for footballs kicked on Earth (solid line) and Mars (dashed line) (p. 44) adapted from 'Association Football on Mars' by Calum James Meredith, David Boulderstone, and Simon Clapton
X-ray of a road-kill green woodpecker and schematic of woodpecker at work (p. 53) from 'A Woodpecker Hammer' by Julian F. V. Vincent, Mehmet Necip Sahinkaya, and W. O'Shea
Geometric division of an elephant from 'Estimation of the Total Surface Area in Indian Elephants (Elephas maximus indicus) (p. 65) by K. P. Sreekumar and G. Nirmalan
Fish's stroke frequency vs. swimming velocity in six platypuses (p. 66) adapted from 'Energetics of Swimming by the Platypus Ornithorhynchus anatinus' by F. E. Fish, R. V. Baudinette, et al.
N.B. Lizardfall plummets in December adapted from 'Arboreal Sprint Failure' (p. 71) by William H. Schlesinger, Johannes M. H. Knops, and Thomas H. Nash
Detail from Egede's monstrous account of 1741 and North Atlantic right whale penis of 2001 (p. 75) from 'Cetaceans, Sex and Sea Serpents' by C. G. M. Paxton, Erik Knatterud, and Sharon L. Hedley. Photograph reproduced by permission of the New England Aquarium, Boston, Massachusetts.
Sheep personality vs. location adapted from 'Effects of Group Size and Personality on Social Foraging' (p. 80) by Pablo Michelena, Angela M. Sibbald, Hans W. Erhard, and James E. McLeod
Distance between two cows during grazing phase (p. 83) adapted from 'How are Distances Between Individuals of Grazing Cows Explained by a Statistical Model?' by Masae Shiyomi
Figure: 'Comparison of amount of faeces accumulated under the lavatory seats in a lavatory at the Antarctic' (p. 96) from 'Review of Medical Researches at the Japanese Station (Syowa Base) in the Antarctic' by H. Yoshimura
Digestive damage to a surviving shrew humerus and surviving shrew tibio-fibula (p. 105) from 'Human Digestive Effects on a Micromammalian Skeleton' by Brian D. Crandall and Peter W. Stahl
How to generate a shock pressure wave for tenderizing an article of food (p. 112) from US Patent no. 3,492,688
Figure: 'The relationship between craving, built, and the eating of chocolate bars …' (p. 115) adapted from 'The Development of the Attitudes to Chocolate Questionnaire' by David Benton, Karen Greenfield, and Michael Morgan
The long and short of bartenders' pours (p. 130) adapted from 'Shape of Glass and Amount of Alcohol Poured' by Brian Wansink and Koert van Ittersum

Fig. 1 of 13… (p. 147) adapted from US Patent no. 7,266,767

Figure: 'Two-way interaction of number of meetings and perceived meeting effectiveness to predict job satisfaction' (p. 156) adapted from 'Not Another Meeting!' by S. G. Rogelberg, D. J. Leach, P. B. Warr, and J. L. Burnfield

The patented Kiss-Throwing Doll (p. 168) from US Patent no. 3,603,029

Figure: Types of mounts, with and without mobile phone (p. 180) adapted from 'Effects of Exposure to a Mobile Phone on Sexual Behavior in Adult Male Rabbit' by Nader Salama, Tomoteru Kishimoto, Hiro-Omi Kanayama, and Susumu Kagawa

Changes in beard growth during a short stay on the island (p. 185) adapted from N.A., 'Effects of Sexual Activity on Beard Growth in Man', *Nature* 226 (30 May 1970): 869-70

Evidence of a huge Parisian tooth-yanker (p. 196). Courtesy of the Wellcome Library, London.

Dr Bean's twenty-year nail chart (p. 203) from 'A Discourse on Nail Growth and Unusual Fingernails' by William B. Bean

Number of deaths from myocardial infarction, French men vs. French women (p. 208) adapted from 'Lower Myocardial Infarction Mortality in French Men the Day France Won the 1998 World Cup of Football' by F. Berthier and F. Boulay

Figure: 'A tractor backhoe using a square section clamping gripper driver to hold, revolve and press a casket into a pre-bored or augered hole' (p. 229) from US Patent no. 8,046,883

The eponymous number of clasp knives (p. 231) from 'Account of a Man Who Lived Ten Years After Having Swallowed a Number of Clasp-Knives, with a Description of the Appearances of the Body After Death' by Alex. Marcet

Figure: 'The knife cut which divides the ham into equal area portions' (p. 236) adapted from 'Green Eggs and Ham' by M. J. Kaiser and S. Hossaien Cheraghi

Facial representation of financial performance (p. 242) adapted from 'Facial Representation of Multivariate Data' by David L. Huff, Vijay Mahajan, and William

The tittle obliquity measurer (p. 246) from International patent application no. PCT/CN2007/003282

The centre point – the intersection of common sense and the Peter Principle… (p. 250) adapted from 'The Peter Principle Revised' by Alessandro Pluchino, Andrea Rapisarda, and Cesare Garofalo

As feature in *Perception* vol. 35's 'Last But Not Least' (p. 256) adapted from 'Eye-witnesses Should Not Do Cryptic Crosswords Prior to Identity Parades' by Michael B. Lewis

From 'Net Trapping System for Capturing a Robber Immediately' (p. 258) from US Patent no. 6,219,959

Underarm stab action' demonstrated (p. 267) from 'An Assessment of Human Performance in Stabbing' by I. Horsfall, P. D. Prosser, C. H. Watson, and S. M. Champion

Os, sampled (p. 270) from 'Quantification of the Shape of Handwritten Characters' by Raymond Marquis, Matthieu Schmittbuhl, William David Mazzella, and Franco Taroni

Note from study: The label 'Hetero' refers to 'heterosexual intimate relationships'… (p. 285) adapted from 'Dude' by Scott F. Kiesling

'Audience members overwhelmingly agree that the man on the left is named "Tim" and the man on the right is named "Bob"' (p. 287) from 'Who Do You Look Like?' by Melissa A. Lea, Robin D. Thomas, Nathan A. Lamkin, and Aaron Bell

ABOUT THE AUTHOR

Marc Abrahams is the editor and co-founder of the science humour magazine *Annals of Improbable Research* and a weekly columnist for the *Guardian*. He is the founder of the Ig Nobel Prizes, which honour achievements that make people laugh, and then think, and which are presented in an annual ceremony at Harvard University. Abrahams and the Ig Nobels have been widely covered by the international media, including the BBC, ABC News, the *New York Times*, *Daily Mail*, *The Times*, *USA Today*, *Wired*, *New Scientist*, *Scientific American*, and *Cocktail Party Physics*. He and his wife, Robin, a columnist for the *Boston Globe*, live in Cambridge, Massachusetts.